Lecture Notes in Economics and Mathematical Systems 585

Robin Pope · Johannes Leitner
Ulrike Leopold-Wildburger

The Knowledge Ahead Approach to Risk

Theory and Experimental Evidence

With 51 Figures and 37 Tables

Springer

Dr. Robin Pope
Experimental Economics Laboratory
Bonn University
Walter-Flex-Straße 3
53113 Bonn
Germany
robin.pope@uni-bonn.de

Dr. Johannes Leitner
Professor Dr. Ulrike Leopold-Wildburger
Department of Statistics and Operations Research
University of Graz
Universitätsstraße 15/E 3
8010 Graz
Austria
JohannesLeitner@gmx.de
ulrike.leopold@uni-graz.at

Library of Congress Control Number: 2006934981

ISBN-10 3-540-38472-3 Springer Berlin Heidelberg New York
ISBN-13 978-3-540-38472-4 Springer Berlin Heidelberg New York

Springer is part of Springer Science+Business Media

springer.com

Production: LE-TEX Jelonek, Schmidt & Vöckler GbR, Leipzig
Cover-design: Erich Kirchner, Heidelberg

SPIN 11838791 88/3100YL - 5 4 3 2 1 0 Printed on acid-free paper

Dedication

To Ken Arrow for his encouragement of our focus on secondary (knowledge-ahead-based) satisfactions through his early recognition of our demonstrations that no rephrasing of expected utility theory can accommodate them, and that these lead to deliberate violations of dominance.

Foreword

The joint work of Robin Pope, Johannes Leitner and Ulrike Leopold-Wildburger presented in this little book makes a strong point for a new approach to decision making under risk. It emphasizes aspects of risk preferences largely neglected in the theoretical and experimental literature. Before the outcome of a risky decision is known a decision maker may feel worries or thrills about what will happen at the end. It is convincingly argued that the anticipation of such "secondary satisfactions" is, and should be, an important influence on the decision. The questionnaire data and the experiments support this view.

The answers of participants in the questionnaires about the reasons for their decision are an important basis for the evaluation of the experiment. The evaluation of these questionnaires has led to impressive findings. Like other commonly used research techniques, the analysis of introspective information in questionnaires is not absolutely reliable. Nevertheless it is an extremely valuable source of insight into motivational forces, insufficiently exploited in economics.

It is not my task to give a preview of the book, but I would like to encourage the reader to discover the remarkable results of an unusual theoretical and experimental approach to decision making under risk.

Königswinter, Germany, August 2006

Reinhard Selten

Preface

In this book we break new ground on why and how people decide on insurance contracts, and offer a few insights on how the context of terrorism may affect such demand. Our findings are suggestive also for decisions under risk in other situations.

The book elucidates what is a risk, and what emotional, planning and financial effects arise from deciding to take risks, rather than avoid them, for instance by taking out insurance. It is for the interested lay person, for government officials and private organizations considering providing insurance. It is particularly designed for economics, business, psychology, philosophy, statistics or operations research students in upper levels of undergraduate study, or doing a Master of Business or Public Administration, or in course-work or thesis stages of a doctoral degree. It is also designed for those doing consulting in the public and private sectors, and those with academic research on choice under risk.

The book provides a new approach to risk. It identifies the fundamental role of time in risk bearing. It identifies that what is distinctive about risk is an anticipated *change in knowledge ahead*. For instance we face a risk if we anticipate that there might or might not be an attack on an asset that we possess. We anticipate going through a period of limited knowledge ahead. This is while we do not know whether there will be an attack and whether we shall lose all. We anticipate in the more distant future, *a change in our knowledge ahead*, since at this later date we shall have learned whether or not there was an attack.

In economics and other disciplines, the demand for insurance contracts is typically estimated via standard rank dependent theories eg EUT, axiomatised Expected Utility Theory, or CPT, Cumulative Prospect Theory. These theories have the characteristic of being widely used, but little understood. We explain in the book these theories. We explain them in ways that will be valuable for students and teachers at both the undergraduate and graduate levels, as also for researchers, policy makers and the interest general public. We trace confusing changes in terminology that have occurred over the years. We elucidate numerous misconceptions about what considerations such theories permit decision makers to incorporate.

We identify restrictions in these theories that tend to be ignored in textbooks. As a consequence these restrictions are largely unknown to academic researchers and government and business consultants employing these traditional ways of estimating insurance demand. One restriction these theories impose on insurees is that the satisfaction a person derives from each possible individual outcome is independent of *knowledge ahead*. This is an extreme restriction when the essence of risk is an anticipated *change in knowledge ahead*! It requires these traditional theories to exclude all knowledge-ahead-based positive and negative satisfactions. *Knowledge-ahead-based* satisfactions include thrills, worry, blame, ability (or inability) to commit and to efficiently inter-temporally allocate consumption.

Omitted sources of satisfactions are of little concern if they are minor and thus rarely alter decisions in commonly encountered choice sets. This book offers evidence that knowledge-ahead-based satisfactions are non trivial, that they influence choices in the risk range commonly encountered in a decision whether to insure or to take the risk and stay uninsured. Thus researchers and advisers on insurance employing traditional risk theories that ignore knowledge-ahead-based satisfactions, are at a disadvantage in understanding well-being and people's demand for insurance.

To include these *knowledge-ahead-based* satisfactions we need to move to a different type of theory, SKAT, the Stages of Knowledge Ahead Theory. This book explains the new theory. It also explains how SKAT solves a famous contradiction that has bedeviled decision theory now for over 60 years - since first uncovered by von Neumann and Morgenstern in the early 1940s. We invite readers into the adventure of moving beyond traditional theories that are static with respect to knowledge ahead into a theory that allows for the anticipated changes in knowledge ahead that each of us face whenever we decide against the safety of insurance and take a risky choice.

We warmly thank the faculty and students in our five sessions for the information that they provided on how people decide on whether to take out insurance. The book has benefited from suggestions and assistance in the analysis given by Rafael Dreyer, Kjell Hausken, Andreas Orland, Angela Meyer, Norman Roberts, Barkley Rosser, Reinhard Selten, Alex Wearing and Bob Whyte. In the book's preparation we warmly thank

Jing Pan and Ali Zaidi for expert assistance.

Finally, we should like to express our sincere thanks to Vice President Dr. Werner A. Müller, to Katharina Wetzel-Vandai and Christiane Beisel from Springer-Verlag for their continuous support and assistance.

Contents

Chapter 1

Introduction

This book is in two parts. Part I explains SKAT, the Stages of Knowledge Ahead Theory, why it is needed, and how it differs from EUT, axiomatised expected utility theory, and most non-EUT theorising. It also addresses potential objections to SKAT and other methodological issues.

Part II offers evidence on the motivations for choice, and thus on whether it is necessary to go beyond traditional EUT and non-EUT theories to SKAT. It also affords evidence on the steps people take in combining different motivations and considerations to arrive at a decision. The evidence is obtained from a probabilistically paid laboratory experiment concerning a decision on whether to play safe and take out insurance, or to play risky and not pay the cost of insurance.

The glossary / list of acronyms in Appendix D furnishes readers a reminder of the denotation of terms that will be new to many readers. The glossary also presents the meaning of terms long used in insurance and the risky choice literature. This short glossary / list of acronyms is also written so as to furnish the reader with an overview of the books' content, and how to move to a more useful, practical and reasonable decision theory.

Part I begins with Chapter 2. This chapter delineates the benefits of specific serious scenarios, and introduces our experimental one. Chapter

3 explains what EUT really is, limiting satisfactions from outcomes to those evaluated "as if certain". It also traces the changes in terminology that came about in 1950 in the meaning of the words "risk aversion". It explains the counter-intuitive new meaning introduced in that year, Marschak (1950), that has confused many scientists since, causing them to misconstrue EUT as more general and more plausible and rational than is the case.

Chapter 4 introduces with examples the terminology of primary and secondary or *knowledge-ahead-based* satisfactions and identifies those excluded under EUT. Chapter 5 introduces the *stages of knowledge* framework required to reach the higher level framework that von Neumann and Morgenstern discerned was required to include secondary satisfactions. Von Neumann and Morgenstern had wished themselves to include secondary satisfactions in their theory, but had been unable to discern the required higher level *stages of knowledge* framework. They had discerned that the higher level was required to demonstrate the pseudo nature of the contradiction that they encountered within the *static* EUT framework. Introducing the higher level of analysis advised by von Neumann and Morgenstern increases the prescriptive and descriptive validity of our findings. Chapter 6 explains material secondary satisfactions that EUT also excludes, and related planning literature.

Chapter 7 addresses misconceptions about and objections to including secondary satisfactions in a serious prescriptive decision theory. One objection is that consideration of secondary satisfactions violates the dominance principle. A benefit of our specific serious application is that it enables us to expose the dominance principle. Scientists have found the dominance principle plausible and rational largely because they have assessed it in an abstract contextless frame. Once the frame is made specific enough, it can be seen to be implausible and irrational. In this chapter we present and explain the dominance principle, within a serious contextualised scenario that enables readers to understand how inappropriate that dominance principle is in any theory of decision making under risk, but most especially in a theory concerning weighty personal or business matters. The chapter also addresses other misconceptions, including the mistaken conjectures that EUT can include emotional and planning secondary (dis)satisfactions by suitably elaborating its outcomes or more fully describing the decision situation. Chapter 8 explains the biases that enter experimental set-ups that ignore secondary satisfactions. Chapter 9 presents SKAT, the Stages of Knowledge Ahead Theory.

Chapter 10 concerns our scientific methodology and potential objections to the conclusions drawn from our use of participant's self-reports. Our discoveries derive from laboratory generated information on how choices are made with a probabilistic receipt of the outcome of that decision. But choices alone tell us nothing about *why* participants chose as they did. To discern why, we require theories with which to analyse the empirical evidence. For discovering pertinent choice theories, we employ what we see as a more objective procedure in that we use participants' self reports to *supplement* the normal procedure. Normal procedures by contrast rely *exclusively* on experimenter's *own* introspection or the introspection of fellow scientists who constructed or advocated a particular decision theory. Reliance on these normal forms of (typically unadmitted) introspection is influenced by (and hence potentially biased by) that scientists *personal* decision theoretic *education.* We address the various objections to going beyond how we - the "objective" scientists "looking down from above" - think the participants decided. We present the case for using in addition to our "ivory tower" insights, insights on what factors participants themselves claim influenced their choices, and what procedures they claim to have used in arriving at a conclusion. This completes Part I.

Part II of the book concerns evidence from our experimental laboratory investigative technique. We first invite participants to choose. Then before they learn the outcome and payoff of their choice, we ask them questions on factors behind both their decisions to avoid risks by insuring, and their decisions in other circumstances, to take risks by not insuring. Their answers offer us hints on how to model actual risky choice procedures - and rather decisive evidence on the advantages of SKAT in both descriptive and prescriptive risky choice modelling.

Chapter 11 gives the full details of the 6 alternative insurance situations faced by our participants, their probabilistic payoffs, their choices, and the roles of context and risk order. Chapter 12 compares decision under dictated choices and menu constructed ones (willingness to pay). It shows how to use the technique devised by Ramsey (1926) and Neumann & Morgenstern (1947) to deduce from these data the shape of the utility function people would have if they obeyed EUT. We show that the weird shapes so derived are incompatible with the widespread assumptions of risk neutrality or constant relative risk aversion, or indeed for nearly over 80%, with any function that accords with the axioms of EUT. Chapter 13 examines the ways in which nearly all participants are influenced in choices by secondary satisfactions. Its findings demonstrate the importance of employing the more rational stages of knowledge ahead framework so as to avoid missing the role of secondary satisfactions in descriptive as well as in prescriptive policy oriented analyses of choice under uncertainty.

Chapter 14 presents our discoveries on how participants utilize information on *different* sources of anticipated satisfactions to choose. The principal ways are incompatible with EUT, and indeed incompatible with any standard rank dependent approach. The principal ways are compatible rather with the algorithmic "production systems" approach of Cyert & March (1963). In turn "production systems" relate to the aspiration adaptation approaches of Simon (1955), Sauermann & Selten (1962) and Selten (1998), the "take the best" approach of of Gigerenzer (2003) and the priorities heuristics of Brandstätter et al. (2006). We find however effects of framing and context that preclude a general algorithm appropriate for all risky choice situations.

Chapter 15 indicates how governments and firms may exploit our findings, whether for their own benefit, that of potential insurees, or for both contracting parties in terrorist and non-terrorist contexts. Chapter 16 summarises what we have discovered about the more rational SKAT, Stages of Knowledge Ahead Theory, and offers directions for further research for discerning and measuring secondary satisfactions in various contexts.

Part I

Theory and Methodology

Chapter 2

A Serious Specific Scenario for a Serious Risky Choice Theory

We always have to abstract and simplify in order to analyse, to theorise. The issue is whether we make *appropriate* abstractions and simplifications in our analysing, theorising. *Appropriate* abstractions and simplifications capture the major cause-effect chains that decision makers do or ought consider.

The choices we consider - the scenarios on which we focus - largely determine how we analyse, theorise. Simple abstract choices are widely perceived as denoting:

1. the universality of the conclusions drawn, and
2. the objectivity and scientific validity of the conclusions drawn.

Theorists and experimentalists alike have on the basis of 1 and 2 favoured being simple and abstract. We show below that perceptions 1 and 2 are erroneous, that usage of abstract choices has damaged theoretical, econometric and experimental work on risky choice. If we make the frame serious, complex and concrete, this aids us in developing our theory of risky choice for serious real world problems, in developing relevant experiments, and in interpreting sensibly field and experimental data, Pope (1983).

Friedman & Savage (1948) advocated that scientists concentrate on simple lotteries to analyse and to theorise about risky choices. Their advice has been followed fulsomely in experimental work. In theoretical work, their advice retains the frivolous social gambling focus out of which EUT arose, namely puzzles in aristocrats playing social games of chance and employing mathematicians to seek solutions to these, Pope (1996/7).

What have been the adverse consequences of the lottery ticket focus? Only a limited number of theorists and experimentalists buy lottery tickets, and even for those who do, it involves a trivial sum and doing an embarrassing thing in the eyes of many colleagues. The net effect is that few lottery ticket purchasing scientists dwell on the purchase for long enough in an unembarrassed enough way to recognize their own spectrum of motivations. When logical positivists focused on trivial ethical issues like posting a letter, they constructed nothing more profound or useful than the boo-hoorah theory of ethics. We can better understand the pros and cons of a decision theory and the implicit biases in an experimental set-up if instead we choose a decision scenario that grips our scientific imagination. This happens when the scenario involves a serious more complex issue that most of us as scientists actually face (Pope, 1983, pp138-9,153).

To illustrate, Savage employed lottery tickets to examine whether he had been mistaken about his own preferences when in 1952 Allais asked him to assess the rationality of expected utility theory in light of Savage's strong preference for making a pair of choices that were incompatible with that theory. Savage stated that although he never intuitively abandoned his original pair of preferences, he had constructed a sure-thing principle that led him to conclude that his intuition was wrong, and that he should have altered his choices. But when the sure-thing principle's procedure for clarifying is examined, it turns out to involve an irrational truncation of probability distributions that renders risky acts sure ones. It turns out that the sure-thing principle can only "clarify" in the designed way, for people who disobey EUT and reap secondary satisfactions from having full knowledge ahead - from being certain, Pope (1991a). It is rather unlikely that Savage would have failed to see the inappropriateness of substituting probability distributions with the risks truncated out, had he stepped outside abstract lottery ticket scenarios and looked at the implications of his sure-thing procedure for one of the serious risky decisions he had faced in his own life. He robbed himself of this by choosing to analyse the dilemma he faced in answering Allais' original lottery questions with an alternative lottery ticket scenario.

Another illustration of the importance for scientists of selecting serious choices is the welter of discoveries in the early 1950s, that few advocates of EUT consider it appropriate to use EUT in life and death situations. This normative repudiation of EUT in life and death situations leads on to the question, when does a situation cease to be sufficiently serious - become sufficiently trivial - that EUT becomes plausible? But when the decision is sufficiently trivial for EUT to be plausible, is not something that has been investigated either normatively or descriptively by those adhering to EUT. Quite a few EUT+ theories are also subject to the problem of being implausible in life and death situations, and for the same reason, developed without an adequate focus on serious complex choices made by the theorists involved.

Widening the range of choices, to include ones that involve serious issues has another advantage. It allows us to understand how seriousness alters the relative importance of different instances of primary and secondary satisfactions, (Pope, 1983, p139), Albers et al. (2000). We shall give examples of these alterations in Chapter 4. Furthermore, some of the major determinants of serious choices play no role in trivial choices, and thus get missed altogether when we confine our theorizing and experiments to the trivial, Pope (1991b, 2001, 2004, 2005a,b).

Even when the monetary payoffs are identical, choices made, and thus what determines choice, varies with framing and context. Thus Berg (1970) found different decisions, based on a different notion of fairness, when payoffs were framed as profits than when framed as net asset position, even though both frames have identical profit, and thus identical net asset position implications. Krzysztofowicz (1983) found that engineers made more conservative choices with respect to risk when the risk concerned others for whom they as water engineers were responsible, than when it concerned their own assets. Albers et al. (2000) found that asset choices were more conservative choices with respect to risk when the context was at exam time and students had an unattractively high level of tension in their lives. Pope (2001) found that riskier or safer group mountaineering ventures were classified as normatively desirable according as the decision making was democratic or autocratic. Hausken (2004) found safer choices in the context of terrorism than other risks involving the same probability distribution. A range of other ways in which context matters is in Harrison & List (2004). Dohmen et al. (2005) provide evidence that the best overall predictor for any specific behaviour is from the corresponding context-specific measure, and that abstract lotteries have limited predictive power for behaviour in specific contexts.

When in an experiment, the context is left blank (supposedly abstract), participants imagine the closest specific context, which, as noted in Conlisk (1993), for most is the frivolous one of lotteries and other social games of chance for teeny stakes. In actual decisions, there is ever a context, and ever a frame. We are claiming false generality if we do not state our frame and context. If we have left these blank, we should therefore be cautious in claiming relevance beyond that of certain social games of chance.

In light of the benefits of a serious actual scenario, the experimental set-ups here employed adopt a reasonably serious decision in a reasonably concrete context, namely as follows. An attack over the next X hours would annihilate a \$100 asset completely. The chooser discovers that he has the opportunity to commit at time $t = 0$ to pay \$15 for full protection from attack of that asset over the next X hours. The insurance contract guarantees the chooser to have an asset of \$100 – \$15=\$85 after the X hours.

Thus in our scenario there are three outcomes after the X hours. After the X hours, the person has either \$0 or \$100, having decided *against* the insurance contract, or the person has \$85 having decided *for* an insurance contract. That is if Y_i denotes the outcome after the X hours, $i=$ \$0 or \$85 or \$100.

Chapter 3

What Expected Utility Theory Really Is: What Its Notion of Risk Aversion Excludes

In our scenario, after the X hours, the outcome Y_i is $i=$ \$0 or \$100 if the person decides on the risky act of remaining uninsured, and \$85 if he decides on the sure act of taking out insurance. EUT has a two-step procedure for determining the value of an act.

3.1 Restrictions on How Outcomes Map into Utilities

Step 1 is to impute values U_i, to an act's outcome(s), respectively U_0 when the outcome is \$0, U_{85} when the outcome is \$85, and U_{100} when the outcome is \$100. The U_i's must map into an inherently uni-variate cardinal scale, unique apart from origin and scale. This means that even if there are several dimensions to the value of an outcome, these must be collapsible into a single dimension mappable into a real number.

Furthermore in arriving at the three U_i's, the EUT chooser must be *independent of knowledge-ahead.* Ie the chooser must exclude consideration of how risky or sure he is at the point of choice about getting that particular Y_i. EUT thus constrains the chooser to put the identical utility number to a particular outcome Y_i regardless of p_i, its probability of occurrence at the point of choice. Ie the outcome Y_i itself (\$0 or \$85 or \$100) is the only determinant of each U_i,

$$U_0 = U_0(\$0) \tag{3.1}$$

$$U_{85} = U_{85}(\$85) \tag{3.2}$$

$$U_{100} = U_{100}(\$100) \tag{3.3}$$

From equations (3.1) to (3.3), U_0, U_{85} and U_{100}, is each independent of:

a) any relevant party's *knowledge of the choice set* embedding them, and thus independently of blame and ridicule (or praise and honour) being entailed in having chosen to insure when one need not, or having not insured when (ex post) it would have been profitable;

b) the chooser's own *degree of knowledge ahead* of which outcome will occur between $t = 0$ and X hours later when he will learn the outcome, and thus independently of any worry or thrills he may feel, or any considerations of ability to commit to repay or to efficiently plan his consumption between the *pre- and post-outcome* periods.

The constraints of the utility mapping from outcomes into utilities being inherently uni-variate and *knowledge-ahead independent* as per equations (3.1) to (3.3) are imposed also by EUT+ theories.

In summary, step 1 for obtaining the three EUT utilities U_0, U_{85} and U_{100} involves three restrictions:

a) the reduction of multiple dimensions of the value of an outcome to a single dimension;

b) the single value dimension being mappable into a real number;

c) the attribution of U_i's to Y_i's independent of knowledge ahead of the Y_i's, and thus for a given Y_i identical no matter the probability at the point of choice of that Y_i occurring.

As we proceed, we shall discover from our experimental findings that all three restrictions on the U_i's are inappropriate.

3.2 Ramsey Versus the Friedman-Savage Version of EUT

There are two ways for an EUT chooser to keep *knowledge-ahead* and thus probabilities out of U_0 and U_{100} as required by equations (3.1) to (3.3). One is EUT as in Ramsey (1926). Such an EUT chooser evaluates each outcome independently of probabilities. The other way is that of Friedman & Savage (1948). This is to ignore the actual probability and thus riskiness of an outcome, and always evaluate it as if it were certain. In the absence of illusory or actual certainty effects, as discussed in the next chapter, the Ramsey and the Friedman and Savage versions of EUT yield identical evaluations and choices. Illusory and actual certainty effects cause the two to differ in the utilities attributed, and in some cases this will alter the choice.[1]

3.3 The Changed Meaning of "Risk Attitude"

From equations (3.1) to (3.3), restriction c) above, what is called "risk attitude" in EUT is, in the language of Friedman & Savage (1948), whether:

(i) the *as if certain* utility function is linear, so called "risk neutral",

(ii) the *as if certain* utility function is concave, so called "risk averse", and

(iii) the *as if certain* utility function is convex, so called "risk loving".

To appreciate the implications of (i) to (iii), note that choosers do not know the outcome, and have only a limited degree of *knowledge ahead* whenever they have merely probabilistic knowledge of the outcomes - whenever the outcomes have non-degenerated probabilities. The characterisation of (i) to (iii) as risk attitude under EUT therefore has nothing to do with a chooser's attitude to differential degrees of *knowledge ahead*, and the impact of having full (as against limited *eg* probabilistic) *knowledge ahead*. What EUT denotes by the term "risk attitude" is thus a utility mapping that excludes all sources of satisfaction pertaining to the chooser's degree of *knowledge ahead*.
What EUT denotes by the term "risk attitude" thus excludes many considerations in risk taking that scientists might intuitively and incorrectly

[1] An example of its altering choice is in Pope (2004).

infer it includes. What EUT denotes by the term "risk attitude" may nevertheless be useful within EUT. But its EUT denotation detracts from scientists' understanding that other essential aspects of risk are now excluded from the words risk attitude. Prior to about 1950 and the widespread adoption of EUT, many other considerations in risk taking were included or includable in the denotation of the term "risk attitude".

The EUT terminology changing the denotation of "risk attitude" to a concave "as if certain" EUT utility mapping was introduced by Marschak (1950), and taken up by Arrow and Pratt in what are now known as the Arrow-Pratt measures of risk aversion. For further details, see Pope (1996/7).

Choice is not between the U_i's, the utilities of the outcomes U_i, but between acts. We therefore need to proceed to step 2 and look at the restrictions that EUT imposes on how an act's U_i's convert into act values.

3.4 Restrictions on the Aggregation of Utilities

Let us use V to denote the evaluation or utility of an act. To the sure act of insuring, the EUT chooser gives the value $V_{insured}$ = U($85). To get $V_{uninsured}$, the value or utility the risky act of staying uninsured, the EUT chooser employs an atemporal rule for *aggregating* the value of the two possible outcomes. He gives U($100) and U($0) each a weight, w_{100} and w_0 respectively, and then adds them up:

$$V_{uninsured} = \sum_{i=0,100} w_i U(Y_i) = w_0 U_0(\$0) + w_{100} U_{100}(\$100) \qquad (3.4)$$

What sort of weights should an EUT chooser use? Should he simply give equal weight to these two mutually exclusive possible utilities U($100) or U($0)? As we shall see later, our experimental results suggest that exceedingly few people use any such weights. There are many weighting rules proposed, with decumulative probability weights (rank dependent theories) frequently advocated. Under EUT, the simplest rank dependent theory, the chooser is constrained to add up U($100) and U($0) each weighted by its probability.

$$V_{uninsured} = \sum_{i=0,100} p_i U(Y_i) = p_0 U_0(\$0) + p_{100} U_{100}(\$100) \qquad (3.5)$$

p_i atemporal aggregation weight for $U(Y_i)$, *outside time,*
$U(Y_i)$ anticipated utility of outcome Y_i in *real time.*

This restriction prevents him from being more conservative than the probability of the bad outcome would indicate, and thus deprives him of this consideration in seeking insurance. It also prevents him from being more attracted to the good outcome and giving it a higher weighting than its probability would indicate, and thus deprives him of this consideration in avoiding insurance.

Under EUT, he sees if from (3.5) the risky act has a higher utility than the sure act of taking out insurance, and selects it if this is the case, is indifferent if both have an identical V, and selects the sure act of insurance otherwise.

In reading EUT's equation (3.5) for the risky act, do not mix up the p_i and the (Y_i) terms. The words written below them are designed to help readers distinguish:

a) the *atemporal* role of probabilities p_0 and p_{100} as weights in *aggregating* the mutually exclusive possible utilities U($0) and U($100) with

b) how each outcome Y_i, respectively $100 and $0, maps into its utility number U(Y_i), respectively U_0 and U_{100}, namely as per equations (3.1) to (3.3), ie independently of knowledge ahead, of probabilities.

Chapter 4

Primary and Secondary Satisfactions

It was partly to help alleviate confusions on what satisfactions EUT excludes, that Pope (2001) introduced the terminology of secondary satisfactions, mentioned in the introduction and its counterpart, primary satisfactions, here explained for the scenario of this book's experiment. For simplicity, we shall start at the point of having to decide whether to insure. In choosing whether to take the insurance offered, experiences *after* choice cannot influence choice, even though they may affect actual welfare. Choice between these acts depends rather on satisfactions under that insurance contract as compared to staying uninsured, *as anticipated at the time of choice.* A chooser may anticipate what positive and negative satisfactions may eventuate if each act were chosen. Decision theories differ in part according to what sources of satisfactions they permit a chooser to consider.

4.1 Primary Satisfactions

Primary satisfactions can be positive or negative (dissatisfactions). They stem from *knowledge ahead independent* sources of satisfactions *as anticipated at the time of choice.* Some primary satisfactions, namely those primary satisfactions occurring after the risk is passed, are included in EUT.[1] These do not include all satisfactions obtained from using the

[1]This is because, as Samuelson (1952a) notes, the EUT outcome flows must begin after all risk is passed. On the inconsistencies this would give rise to in axiomatisations were the

funds remaining after the attack, either the \$0 or the \$100 if uninsured, or the \$85, the amount remaining after insured if insurance were chosen. They include only those satisfactions reaped from the outcome independently of what the person knew at the point of choice about what would be his future outcome. Ie they include only those independent of his knowing at the point of choice whether that outcome is guaranteed or merely possible.

If as EUT and EUT+ theories do, we limit ourselves to:

(i) the sub-set of primary satisfactions that will be reaped after all risk is passed,

(ii) cases where the utility scale and post risk outcomes are both inherently univariate - not with irreducible multiple dimensions, and

then utility from this sub-set of primary satisfactions is itself univariate and can be mapped in a plane against these particular univariate sorts of post-risk outcomes. The utility shape is concave if there is diminishing marginal utility from primary satisfactions, linear if there is linear marginal utility from primary satisfactions, and convex if there is increasing marginal utility from primary satisfactions. This was perhaps first so depicted in the famous Friedman & Savage (1948) diagrams. In that paper, primary satisfactions are what Friedman and Savage term the EUT utilities, being unaware that there might be certainty effects captured by their version of EUT, and recognizing that EUT excludes from its utilities sources of satisfactions based on a *limited degree of knowledge ahead.*

4.2 Secondary Satisfactions

Secondary satisfactions are the counterpart to primary satisfactions, the complementary class of satisfactions from primary ones. Secondary satisfactions can be positive or negative. They stem from *knowledge ahead based* sources of satisfactions *as anticipated at the time of choice.* They are termed secondary since they derive from primary satisfactions, not because they are necessarily less important in insurance and other decisions. Since in the case of secondary satisfactions, utility derives not

flow to start earlier, see Pope (2005b).

simply from the outcome, but on its riskiness or certainty, it is infeasible to trace out the secondary satisfactions function on a plane with outcomes on the other axis. Multiple other dimensions (axes) will in general be needed to capture the various aspects of risk whose bound at one end is certainty.

Secondary satisfactions stemming from a limited degree of knowledge ahead are excluded under EUT, and EUT+. Whether secondary satisfactions stemming from certainty - full knowledge ahead - are excluded from EUT, depends on the version of EUT. These are excluded under the Ramsey version, included (also illusory ones) under the Friedman and Savage version.

4.3 Earlier Terminologies

There have been many prior terminologies for primary and secondary satisfactions. For an extensive account, see Pope (1996/7), of which a few are listed in Table 4.1.

Table 4.1: A Bird's Eye History of Terms for Primary and Secondary Satisfactions

Date	*Author*	*Primary Satisfactions* (from outcome Y_i)	*Secondary Satisfactions* (from the uncertainty)
1920	Marshall	utility of money	utility of gambling
1926	Ramsey	ethically neutral propositions	ethically non-neutral propositions
1947	Neumann-Morgenstern	utility	utility of the mere act of taking a chance
1950	Marschak	risk neutral, risk averse, risk loving, each as a shape of the EUT "utility function [mapping Y_i into U_i] of sure prospects"*	no names because excluded under EUT as inappropriate, ones that "We treat in the same way [as] ... mistakes of arithmetic ... [and] logic"*
1952	Allais	psychological distortion of the monetary values	inherent liking or disliking of risk
1972	Marschak-Radner	ones independent of beliefs	ones dependent on beliefs
1978	Harsanyi	utility	process utility
1988	Hammond	consequentialist utility	utility from irrational sources
1995	Pope	utility of the outcome	utility of chance
1998	Pope	utility of the outcome	attractions to and repulsions from chance

* words in quotation marks are from (Marschak, 1950, pp120, 139-140)

Each prior term for secondary satisfactions had disadvantages. In the case of Marshall's "utility of gambling", these are as three.

1. It has enticed researchers to focus on addictive gambling instances and miss the overall significance of this class of satisfactions: substituting "chance" for "gambling" as in Pope (1995) mitigates but does not eliminate this misleadingly frivolous and largely negative connotation.
2. The word utility connotes the index of axiomatised expected utility theory which in fact is constructed excluding secondary satisfactions.
3. It is too easily mistakenly construed as meaning something else, namely the effect of simple probability weights of outcomes on act choice in the case of people with diminishing marginal utility. Indeed Friedman & Savage (1948) proposed that its meaning be changed from the Marshallian usage to mean precisely this, and the change in terminology has partially taken place as they proposed.[2]

Ramsey's terms for primary and secondary satisfactions are easily misconstrued as a distinction between moral and immoral satisfactions. In von Neumann and Morgenstern's utility of the mere act of taking a chance, readers tend to miss the significance of the word "mere", and since there is no corresponding phrase for the class of primary satisfactions, and misconstrue it as including all satisfactions.

From Marschak's 1950 terminology discussed in section 3.3 - there has been widespread confusion with many scientists falsely believing that the three words "risk neutral", "risk averse" and "risk loving" refer to secondary satisfactions. But in fact, under Marschak's change in terminology, these three words refer exclusively to a sub-set of primary satisfactions. Because of this confusion, we shall largely refrain from using these three words in the rest of the book.

In Marschak and Radner's 1972 terminology, some readers mistakenly think that tastes dependent on beliefs just means that the decision maker uses atemporal probabilities as weights in aggregating the mutually exclusive utilities as in EUT's equation (3.5) of chapter 3. Harsanyi's

[2] On the distinction between risk avoidance because of (i) diminishing marginal utility, (ii) conservative probability weightings of outcomes, and (iii) secondary dissatisfactions, see Pope (1995).

coinage of process utilities for secondary satisfactions gets confused with satisfactions from the act itself which are independent of the outcomes set and their associated likelihoods.[3] Indeed the word process is currently used both to denote such act satisfactions as well as what are here termed secondary satisfactions, *eg* in Menstrel (1999).[4] Hammond's term consequentialist for primary satisfactions is misleading. Secondary satisfactions also have consequences for people. Denying this readily leads to Hammond's position that considering anything other than primary satisfactions is irrational because one is going beyond consequences.[5] The new terminology of primary and secondary satisfactions is designed to generate a clearer grasp of how the class of satisfactions partition, and aid understanding of the import of secondary satisfactions in serious reasonable / rational decision making. This has been an uphill battle since the whole class of secondary satisfactions is besmirched with a negative, frivolous name.

4.4 Emotional Secondary Satisfactions in Insurance

Vickrey (1945) claimed that insurance for items of moderate but not huge worth (such as jewellery) to avoid a negative secondary satisfaction, worry. In related contexts evidence of the importance of worry and anxiety in choices taken is provided from introspective data eg Bell (1981), and from experimental and epidemiological data, eg Sunstein (2003), Schade et al. (2004), Mellers et al. (1997, 1999, 2001). In other contexts, people fail to take precautions (to insure) in order to have anticipated thrills. They prefer a risky fair bet over the safe amount to add tension, Albers et al. (2000). If a person fails to be daring, runs for insurance, he may anticipate a risk of the negative secondary satisfaction of being blamed for timidity (see Canaan (1926, reprinted 1963)). Indeed Hagen (1985) reports a case of a chief executive office being fired for not

[3]For what is perhaps the first axiomatised model of what are here termed satisfactions from the act, see Luce & Marley (forthcoming). For a recent axiomatisation of what are here termed secondary satisfactions, see Diecidue et al. (1999).

[4]The sources of many such satisfactions from the act itself can be included in an extended notion of an outcome without destroying the axiomatic basis of the expected utility procedure since they can be specified independently of the likelihoods of the outcomes. This is not the case with attempts to include secondary satisfactions through an extended notion of an outcome. See chapter 7.

[5]For critiques of the unduly narrow interpretation of consequences found in defences of expected utility theory such as Hammond's, see eg McClennen (1988, 1990), Pope (1991b, 1995), Munier (1996).

taking a related risk, namely to develop a North Sea oil field offered him by the Norwegian government. Later the stockholders discovered what could not be known at the point of choice with *limited knowledge ahead*, that it would have been an extraordinarily profitable oil field to develop. Hagen argues that the chief executive officer erred in not anticipating this risk of the negative secondary satisfaction of being blamed.

4.5 Emotional Secondary Satisfactions in Other Contexts

There are also numerous observations of emotional secondary satisfactions in non-insurance contexts. We shall mention here only a few - to help readers relate this class of satisfactions to other literature. The philosopher Plato in explaining scholarship (1967), the Adam Smith in his inquiry into the origins of scientific discoveries (1778), Charles Darwin in his observations of animals (1876), the trade theorist Edward Canaan (1926), all give a leading role to curiosity, a positive secondary satisfaction from limited knowledge ahead allowing discoveries. Pascal (1670) describes people being bored if they knew how much they would win at the beginning of a day of social gambling, ie describes a negative satisfaction from *full knowledge ahead*, a positive secondary satisfaction from *limited knowledge ahead*. Ramsey (1926) describes love and hatred of excitement, positive and negative secondary satisfactions from limited knowledge ahead. Von Neumann, Savage (in his pre EUT days), and more recently Bell (1981) and Loomes & Sugden (1982) describe and model regret, a negative secondary satisfaction at having chosen the act from *limited knowledge ahead* if later the chosen act turns out badly. Bell (1981, 1985) describes and models disappointment, a negative secondary satisfaction, from *limited knowledge ahead* that the chosen act's outcome was worse than it might have been. Hart (1930) describes a cleaning lady buying a lottery ticket to buy hope, for dreaming until the ticket was drawn, so that limited knowledge ahead allowing hope, gave her a positive secondary satisfaction. Caplin & Leahy (2001) describe the opposite, anxiety of what will happen, the negative secondary satisfaction from limited knowledge ahead, and how its anticipation alters choice. In a related vein, Bosman & Winden (2005) identified anxiety as a cause of reduced investment in risky projects in situations where neither EUT nor cumulative prospect theory could explain the lower investment levels. Loewenstein (1987) provided evidence of situations in which having full knowledge ahead (certainty) for a longer period of time is attractive, a positive secondary satisfaction. Dufwenberg (2002) offers as an instance of a negative secondary satisfaction guilt at the spouse's trust in a case of *limited knowledge ahead* turned out to be trust misplaced. Janis & Mann (1977), Lazarus (2000), Somerfield & McCrae (2000) and Hardie (2002) analyse stress in choosing, mainly instances of this being unpleasant, ie of negative secondary satisfactions from limited knowledge ahead of what would be a good decision. Grant et al. (1998), Hoel et al. (forthcoming) describe people who prefer to be ignorant on whether they may contact a disease, ie a positive secondary satisfaction from having limited knowledge ahead in this respect, and Wu (1991) related preferences of some for ignorance. Markowitz (1959), (chapter 10) and Berns et al. (2001) analyse the pleasures of surprise, a positive secondary satisfaction from previously having had limited knowledge ahead.

Scitovsky (1976) amassed evidence from animal and human experiments on people needing to find situations in which they have limited knowledge ahead, the possibility of achieving something, of undertaking challenges and enjoying a tiny bit of fear. If everything they do is known ahead, he showed, people get bored. When lacking such challenges, they engage in activities like watching a horror movies or on roller coasters that yield them this missing and desired whiff of fear. He advocated education to help people see and undertake productive challenges and avoid boredom.

Chapter 5

The von Neumann-Morgenstern Contradiction When Including Secondary Satisfactions Solved with Stages of Knowledge

Von Neumann and Morgenstern added in the 1947 edition of their famous book axiomatising EUT an appendix to deflect criticism that they had omitted secondary satisfactions and thereby kept to an excessively narrow risky choice theory. In this appendix they explained that their original intent had been to construct a broader decision model than EUT, but pointed out (using the language of quantum physics) that secondary satisfactions involve a complementarity (or as we would say an interdependence) between mutually exclusive outcomes. For instance a positive secondary satisfaction such as an enjoyable thrill is happily feeling the tension between the two or more possible outcomes. But how can there be this tension, this interdependence, von Neumann and Morgenstern asked, when the outcomes are mutually exclusive? Such interdependence they said is a "contradiction" on the "level" of our theoretical framework. Von Neumann and Morgenstern stated that it was such a "deep" problem to solve the contradiction and include the "elusive" secondary satisfactions, that they had to leave the task to future researchers (Neumann & Morgenstern, 1947, pp628-32). Near the end of his life Morgenstern reiterated this joint desire of him and von Neumann, and their regret that the task had still not been accomplished (see Morgenstern (1974, 1979)).

Von Neumann and Morgenstern were correct in discerning that the contradiction was one of perception only, a problem of their framework not being at the right level for risky choice. What is required for risky choice is a minimum of a two period future at $t = 0$, the time of contract choice. The future must be divided at $t = 0$ into (i) a period before it is anticipated that the risk will be resolved, and (ii) a period after it is anticipated that the risk will be resolved. Ie for a choice to be risky, the future from $t = 0$ must be divided at least into (i) and (ii). Term (i) a pre-outcome period, namely a period of *limited (merely probabilistic) knowledge ahead* that lasts from $t = 0$ until the time $t = k$. At time $t = k$, $k > 0$, the risk about the outcome will be dissolved. Term the time from time $t = k$, the *post-outcome* period of *full knowledge ahead*, of no uncertainty about the outcome, Pope (1983). Von Neumann and Morgenstern had sought only to axiomatise and model the simplest risky choice problems. But even for these there must be this minimum two period future. Without the *pre-outcome* period with the anticipated change in knowledge ahead at time $t = k$, there can be no risk.

The error in their framework had been to overlook this *pre-outcome* period, and thus include a false simultaneity postulate in their axioms. This error introduces the contradiction that the outcomes of sure contracts (those with the outcome known at $t = 0$), and risky contracts, those with the outcome unknown at this time, had their outcomes known simultaneously, Pope (1985). The error of omitting the *pre-outcome* period creates a corresponding timing fallacy in the definition of risky acts in EUT under which as in *eg* Harsanyi (1978), these are inconsistently defined as "probability mixes of sure acts". See Figure 5.1.

What therefore is missing from EUT and most theories is this division of the future into a risky immediate *pre-outcome period* of merely *probabilistic knowledge ahead.* All that EUT includes implicitly is the right hand side of Figure 5.1, that is the more distant future of the part of the future, the *post-outcome* period when all the risk will be past, when both will have *full knowledge ahead.* As von Neumann and Morgenstern had noted and Samuelson (1952a) reiterated, EUT is for a situation when the risk is resolved in the very near future, and took it for granted that the intervening instant beginning with choice before that resolution of risk is irrelevant to the chooser.

Great advances have been made in the realm of generalising or of-

Figure 5.1: EUT's Jump Through of the Period of Risk to Certainty

$t=0$ choice	$0 \leq t < k$ *Pre-outcome period* of risk, of limited knowledge ahead	$t \geq k$ *Post-outcome period* of certainty, of full knowledge ahead				$t=0$ anticipated utility in the post-outcome period
	This risky period does not exist so it is impossible to have thrills from the risk or worry about which outcome will occur or to suffer from being unable to commit on a loan as unsure whether can repay or to have a dinner with a friend as unsure whether will be wealthy enough to have warranted it	possible outcomes		Y_i Knowledge-ahead-independent sources of utility		$U(Y_i)$ utilities
Risky Act of not Paying for Protection: \$0 or \$100		\$0	$\longrightarrow$	Y_0	$\longrightarrow$	$U_0(Y_0)$
		\$100	$\longrightarrow$	Y_{100}	$\longrightarrow$	$U_{100}(Y_{100})$

fering alternative non-nested more realistic substitutes for EUT. This experiment addresses a factor omitted in nearly all of these, the risky *pre-outcome* period, during which the chooser anticipates having only limited knowledge ahead of the outcome. The chooser has leapt over the risk and landed in certainty, *eg* the moments theory of Coombs & Huang (1970), the prospect theories of Kahneman & Tversky (1979); Kahneman et al. (1993), the disappointment and regret theories of Bell (1981, 1985), Loomes & Sugden (1982), the anticipated utility of Quiggin (1982, 1993), the aspiration-rank dependent theory of Lopes (1984), the prominence theory of Albers (2001), the priorities theory of Brandstätter et al. (2006).

To consistently include secondary satisfactions free of contradiction, decision theories - even non EUT ones - must shift to a minimum of a two-period future, with the second more distant period in the future beginning when at least one (and often each) party anticipates experiencing a change in knowledge ahead from merely probabilistic to full knowledge. Figure 5.1 illustrates the pseudo contradiction created by leaving out the *pre-outcome* period. In more complicated situations, limited *knowledge ahead* may not be in the form of numerical probabilities, there may be multiple anticipated *changes in the degree of knowledge ahead*, and thus multiple *pre-outcome* periods, and *changes in degree of knowledge ahead* may precede rather than coincide with changes in some of money that the parties receive.

Further some forms of secondary satisfactions involve *earlier changes in knowledge ahead* in the decision process. In this case we need to consider a series of time points $t = \ldots - 1, 0, k$ at which there is a *change in knowledge ahead.* For instance, to suffer regret, or to get blamed or ridiculed for a bad outcome (such as having no money after a terrorist attack) is a consequence of an even earlier *change* in *knowledge ahead.* It is a consequence of the fact that there used to be the alternative for such persons of paying for protection, and that they chose not to do so. Such negative secondary satisfactions as blame (which as Hagen (1985) observed causes chief executive officers to be fired), arises out of there having been an earlier *change in knowledge ahead*, that from the *pre-contract decision* situation, to a choice made on whether or not to buy insurance. Sauermann & Selten (1962), Simon (1991b), Selten (1998) emphasise implicitly even earlier *changes in knowledge ahead.* These are from not knowing any available contracts to having negotiated or discovered one, and then if the search continues, from having negotiated or discovered only one, to having discovered or negotiated a second, and so forth. Such *changes* as more is learned about the choice set will also be relevant to our policy advice, namely the issue of the time and expenditure that governments and other insurers should expend in exploring and negotiating to create situations and associated contracts.

In Figure 5.2 and Tables 5.1 and 5.2 we illustrate for our scenario and the case of common knowledge and thus a shared epistemic division of the future, meaning a shared division of the future demarcated by stages of knowledge ahead. We include three epistemic periods that von Neumann and Morgenstern (and EUT and even virtually every disappointment, and regret theory) omitted:

Period 1 the period prior to identifying the choice set,
Period 2 the period prior to choosing an act within the choice set, and
Period 3 the period prior to learning the outcome of the chosen act.

Each of these periods ends with a *change in knowledge ahead* about a different matter. All that von Neumann and Morgenstern and EUT include is period 4, the period that begins upon learning the outcome of the chosen act (and that lasts indefinitely in our simple scenario, not being followed by any other change in knowledge ahead).

In Table 5.1, period 1 is when the parties explore and negotiate on contracts are feasible. It ends with a discovery, *ie* the first change in knowledge ahead, ΔK_1. This first change is caused by acquisition of the knowledge of the choice set, as comprising only two acts, commit to an insurance contract, or decide on no contract. Period 2 begins with the second change in knowledge ahead, ΔK_2, and occurs at time point $t = 0$ when potential insurees decide whether to commit themselves to this contract, as a consequence now know their chosen act. If they chose the risky act of no contract, they face a period 3, the pre-outcome period of *limited degree of knowledge ahead* of the outcome until their third change in knowledge ahead, ΔK_3, the time when they learn the outcome and enter the *post-outcome period* of full knowledge ahead, period 4.

Including these three earlier periods that von Neumann and Morgenstern and EUT omit does two things. First, it dissolves the von

Figure 5.2: The Future Divided Epistemically - by *Changes in Knowledge Ahead*

P_i epistemic periods	P_1 *pre-* choice set	P_2 *pre-* choice	P_3 *pre-* outcome	P_4 *post-* outcome
	$-2 \leq t < -1$	$-1 \leq t < 0$	$0 \leq t < k$	$t \geq k$

time points t

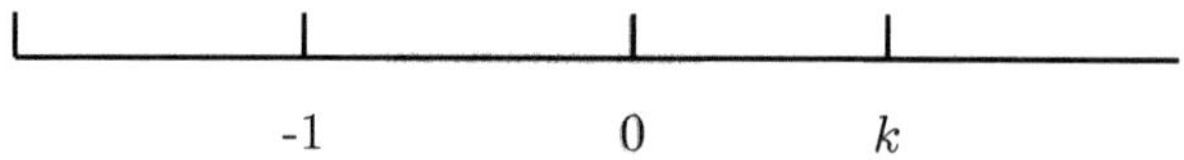

Note:

1 The name of the first three periods starts with "pre-" and then states an aspect of the future that the person will not yet have learned, namely the choice set in P_1, whether to pay for protection and take the insurance contract in P_2, and what will be the outcome of his decision in P_3. P_3 is degenerated, of zero duration, if a person decides to pay for protection, as then he knows at his point of decision $t = 0$ the outcome, ie in this case $t = 0 = k$. If instead he decides to take the risk and not insure, P_3 is, like the other periods, of positive duration, ie $t = k > 0$. The last period, P_4, starts with "post", and then states the aspect of the future the person that the person by the will have learned the outcome.

2 Each time point denotes a *change in knowledge ahead*, with one previously unknown aspect having become known, and the start of a new epistemic period that last until the next change in knowledge ahead.

3 In the case of choosing a sure act, eg deciding to pay for protection in our scenario, the outcome of the chosen act is already known. Thus for sure acts the time points $t = 0$ and $t = k$ coincide - are simultaneous - and the duration of the pre-outcome period is zero. It is only for risky acts that the time points $t = 0$ and $t = k$ fail to coincide - are non-simultaneous - so that the pre-outcome period of positive duration. The false simultaneity postulate in von Neumann and Morgenstern (1947,1972), leading to their contradiction, had two prongs. Prong a) was to explicitly set k to be simultaneous for all acts in their explanation of their axiomatisation. Such simultaneity is incompatible with the theory including both sure and risky acts, when these are distinguished from each other by whether there is full or merely limited knowledge ahead of the outcome at $t = 0$. Prong b) was to implicitly set $k = 0$ for both risky and sure acts in the atemporal axioms (ones lacking any epistemic division of the future from the point of choice).

Neumann and Morgenstern contradiction which arose from omitting the pre-outcome period, Figure 5.1. Second, it dissolves related contradictions concerning secondary satisfactions during a succession of epistemic periods that is during a succession of periods each demarcated from the earlier one by an anticipated change in *knowledge ahead.*

In evaluating a choice of not committing to pay for protection, potential insurees anticipate facing a *pre-outcome* period during which the mutually exclusive outcomes (of no attack or an attack) interact in their brains. They may anticipate that during this *pre-outcome* period, this *limited degree of knowledge ahead* of the outcome will create negative secondary satisfactions of a) worrying whether or not an attack will occur, and b) being unable to commit to spend the money on things desired during this pre-outcome period. That is the von Neumann and Morgenstern contradiction arose from confusing mental events in the earlier omitted pre-outcome period with the later post-outcome period, when indeed only one of the two mutually exclusive outcomes will have occurred.

Even in the more distant part of the future, the *post-outcome period* when they will know whether there was an attack or not, potential insurees may anticipate secondary satisfactions from the historical legacies from prior anticipated changes in knowledge ahead. Eg potential insurees may anticipate that if they do not insure and later an attack occurs they may suffer from negative secondary satisfactions in the *post-outcome period.* These negative secondary satisfactions arise from the fact that there used to be a *pre-outcome* period and indeed a *pre-contract decision period* when they could have chosen protection. They may thus anticipate that after the attack occurs in the *post-outcome period*, accusers can look back to the contract option and pour blame and ridicule on the person who earlier could have taken out insurance because of the previously possible mutually exclusive outcomes of choosing and not choosing a contract. Such ridicule and blame is infeasible, in contradiction with the facts, if there never was such a choice, never was such a *pre-choice* period in which to choose.

What the insurance decision does is to render all such negative satisfactions infeasible to anticipate. For what insurance does is alter by X hours the timing of when all risk is resolved, from X hours later, to the moment when the decision to insure is taken. Thereby insurance eliminates the pre-outcome period and all the positive and negative secondary satisfactions of X hours of *limited knowledge ahead* arising out of the decision not to insure.

In evaluating the alternative choice, the sure act of committing to pay for protection, worry in the *pre-outcome* period is impossible, a contradiction in terms, as there never will be a *pre-outcome* period, of positive duration. Instead the post-outcome period of *full knowledge ahead* begins at $t = k$ precluding the anticipated negative secondary satisfactions listed in Table 5.2 under the risky act of not committing to pay for protection. Under the choice of the sure act, of committing to pay for protection, the chooser can anticipate positive secondary satisfactions of being able to commit to repay a loan and of planning efficiently, including knowing that there will be sufficient funds at the point of choice, $t = 0$ to taking a friend out to dinner.

From Table 5.2, it can also be seen, that when the chooser has to discover relevant available acts, eg having to nominate what amount she would be willing to commit to pay for protection, she may be in a different mindset with the stress and challenge of needing to nominate, than in a situation in which artificially, the choice set is handed to her "on a plate" as a take-it or leave-it ultimatum from the experimenter. The ultimatum deprives her of the activities with associated costs and benefits of the pre-choice set period. It deprives her of having to explore and negotiate, activities that as Simon (1991b) noted, use up a large proportion of the time of businessmen.

Consider now the effects of knowledge and anticipated changes in knowledge stemming from the uncertainties prevailing prior to the point of choice. If the chooser decides against the available protection contract, she can anticipate being blamed and ridiculed if an attack occurs. Being blamed and ridiculed are adverse events that he cannot anticipate if there never was a *pre-decision* period containing the opportunity to commit to pay for protection. In the pre-decision period he may also experience a great deal of stress in determining the appropriate decision, Janis & Mann (1977), information and complexity overload, Omodei et al. (forthcoming), Pope et al. (forthcoming). Decision making stress, information and complexity overload are impossible if there never were a period before the decision is made.

Table 5.1: Three Anticipated Changes in Knowledge Ahead $\Delta K_1, \Delta K_2$ and ΔK_3 After Identifying the Risk of an Attack

Change	*Time*		*Knowledge with Respect to a Future or Past Event*
ΔK_1	from	before the choice set is identified	*Period* 1: only probabilistic knowledge of what the sub-acts search/negotiation in ascertaining and creating available contracts the insurers will identify as the feasible contract.
	to	contract, and risks without it, identified	*New epistemic period, Period 2 starts:* now know with a probability of 1 the choice set, ie sole contract available and what are the risks for those with property at risk who choose against the contract for protection against an attack.
ΔK_2	from	*pre-contract decision*	*Period* 2: only probabilistic knowledge about what the sub-acts of evaluating the contract will identify to each potential insuree on whether or not he or she will choose that contract, ie pay for protection.
	to	contract or not decided	*New epistemic period, Period 3 starts:* Parties know with a probability of 1 who chose the contract.
ΔK_3	from	*pre-outcome*	*Period* 3: lasting for X hours in the case of those with property at risk who rejected the contract - choose not to pay for protection - only probabilistic knowledge of which outcome will be later learned to be the chosen act's *actual* outcome, whether *ie* the outcome will be \$100 or nothing For those vulnerable to an attack who choose the contract - to pay for protection - this period is degenerated, of zero duration.
	to	*post-outcome*	*New epistemic period, Period 4 starts:* parties know with with a probability of 1 the chosen act's *actual* outcome. For those whose property was liable to attack and who rejected the contract - choose not to pay for protection - the outcome is either \$100 or nothing. For those whose property was liable to attack who chose the contract - to pay for protection - this period starts sooner, at $t = 0$, when they already know the outcome, namely \$85, \$100 minus the \$15 cost of protection to which they committed themselves.

Table 5.2: Changing Knowledge Demarcating Eleven Sources of Secondary Satisfactions for the Uninsured

Problem at time $t = -2$		*Anticipated Future*
⇑	1	Bad Stress eg time lost, expense of identifying relevant available acts
Pre-Choice Set period		**and**
(*Available Acts unknown*)	2	Good Stress eg brain growth from challenge of identifying relevant available acts
⇓		
Set identified at time $t = -1$		
⇑	3	Bad Stress eg time lost, expense of choosing
Pre-Decision period		**and**
(*Act unknown*)	4	Good Stress eg brain growth from challenge of choosing
⇓		
decision at time $t = 0$		
	5	Fear, worry that one of the act's possible bad outcomes may come,
		and
	6	Enjoyable Thrills as contemplate the alternative good and bad outcomes that may ensue
⇑		**and**
Pre-Outcome period	7	Disadvantage of being unable to commit and thus enjoy the benefits of commitment
(*Act's outcome unknown*)		such as ability to commit at $t = 0$ to repay a loan for a meal tonight
⇓		**and**
	8	Disadvantage of being unable to plan efficiently, since do not know
		whether future asset level would merit taking a friend out to dinner tonight
		- this future asset level will only become known next period
Learn outcome at time $t = k$		
	9	Blame and Ridicule and Regret since rejected acts had better outcomes,
⇑		**or**
Post-Outcome period	10	Praise and Admiration and Elation since rejected acts had worse outcomes
(*Act's outcome known*)		**and**
⇓	11	Historical legacies, emotional and financial of the prior uncertainty

New problem at time $t > k$
1-8 help in this next pre-decision period, in making the next choice at $t = 3$.
We reason, choose partly through the emotions 1, 2, 3, 6 and 7. For instance brain damage removing the scope for 2 (fear) or for 7 (regret) removes good decision making, Damasio (1984), Camille et al. (2004).

Chapter 6

Material Secondary Satisfactions and Planning

From Figure 5.1 and Table 5.1 it can be seen that during the *pre-outcome* period, the chooser of a risky act may also have material consequences of having only a limited *degree of knowledge ahead* of whether his whole \$100 asset will be annihilated, consequences that EUT excludes. He might for instance feel that he could not borrow money for an evening meal during this uncertain interim if he could only repay in the absence of an attack. He would also have other planning problems on how to allocate his consumption over the next X hours and the more distant part of his future when he will know the outcome, which we may term the *outcome period.* He may have enough cash to invite now a friend for dinner and would like this very much. But if after X hours have passed, he learned that there had been an attack, then he would be so poor that he would rather not have issued the invitation. Recognising this possibility, he decides to refrain. He decides to refrain from issuing the invitation, even though he realises that if it later turns out that there was no attack, his decision will have been inter-temporally suboptimal. He will have excess funds in the later period when it is too late to issue the invitation.

These material negative secondary satisfactions of inability to commit and suboptimal inter-temporal resource allocation - and indeed all other planning difficulties - are non-existent unless there is a pre-outcome period. The impossibility of EUT including them springs from its ignoring the pre-outcome period as depicted in Figure 5.1. That is the impossibility springs from EUT's limitation that its outcomes streams (to which choosers attach utilities) begin only in the post-outcome period - begin after all risk is passed, as Samuelson (1952b) observed. When this axiomatic constraint on when the outcome flow begins under EUT is forgotten, there is introduced the contradiction of acts being simultaneously sure and risky.[1]

6.1 The Flawed "Primitive" Re-definition of the EUT Outcomes Space

When scientists overlook the axiomatic constraint that the EUT outcomes flow must begin after all risk is passed, they begin what they mistakenly construe as an EUT outcome flow that begins in the pre-outcome period. Some notice the obviously implausible decisions under EUT in the face of consumption planning problems with risky income (asset, endowment streams). Some of these have proposed redefining or elaborating outcomes (or equivalently giving a fuller description of the decision situation) to enable EUT to give sensible answers - to include material secondary satisfactions. They have proposed switching the definition of outcomes from one over income (endowment or assets), to one over "end-point" consumption bundles. This is termed redefining the space from one over "induced" to one over "primitive" preferences, or equivalently, giving a "fuller description of the whole decision situation".

On inspection these re-definition proposals turn out to be as invalid, as fraught with contradictions, as the proposals to include emotional secondary satisfactions by redefining EUT outcomes. There are two sequential versions of the re-definition proposal. Under version 1 in the sequence, we take a risky act's outcomes space in terms of an income stream, see what would be the (best) stream of "primitives" - of what ultimately matters to the chooser, typically simplified to a consumption stream - that could be attained from this risky act given all we know from a "full description of the decision situation" under each of its possi-

[1] For an example of how this contradiction can play out, see the contingent loans examples in a decision context, Pope (1991b) and in a game theoretic context, (Pope, 2005b, pp308-320). Contingent loans are received by the borrower before she or the lender know whether they will be repaid, ie in the pre-outcome period. Within EUT, treating these as either part of an outcome stream that begins in the pre-outcome period, or in other ways seeking to take them into account, is infeasible.

ble outcomes, and re-define that act in terms of a probability distribution over these derived consumption streams. Then, a "correct" value is put on this risky act, correct in the sense that the planning difficulties under the act with all the information of a "full description of the decision situation" have been addressed.

This is sensible, but is it within the axioms of EUT? The answer is no. Axioms yielding the EUT property of equation (3.5) are very strong, ie exclude much. While, various alternative axiom sets yield equation (3.5), all have the characteristic that they directly or by logical implication require rankings of a pair of acts to be invariant when the preferred act is replaced by a probability mix of itself and the less preferred act. This condition will not hold generally for our redefined outcomes space for so-called "primitive preferences" any more than it held for the outcomes space for so-called "induced preferences". It will not hold because for some act pairs and some technological / financial inter-temporal transfer options over consumption streams, the mix will yield a reverse preference. Wold's famous *Econometrica* exchange illustrates the sensible consumer's use of his consumer durables to inter-temporally improve his resource allocation under risk by violating EUT, Wold (1952). Previously, the virtue of EUT had been seen in its repeated use, (Marschak, 1950, p139). The rejoinders to Wold, Savage (1952a, 1952b) and Samuelson (1952b) argued the reverse, that EUT should only be applied to one decision in a whole lifetime, admitting that it gave implausible answers for planning under risk if there were multiple decisions over a life. But Pope (1991b) and (Pope, 2005b, p310) furnishes examples demonstrating that the problem for EUT recurs even within the constraint of a person who only makes one decision in the whole of his life.

Confronted with such demonstrations that the induced to primitive redefinitions render the axioms false even for a single lifetime decision, there is a robust oral tradition that the re-definition has been misunderstood. The second version of the redefinition / elaborations proposal is to allege that it is illegitimate for the chooser to consider the primitive consumption streams as inter-temporally malleable. It is asserted that these are end-point consumption bundles, end-point meaning that earlier consumption bundles in the stream are unable to be stored / saved / lent and that later consumption bundles are unable to be used as collateral to borrow etc. But this is a matter of trying to have one's cake and eat it too. If the outcome streams denoting outcomes in EUT axiomatisations are inter-temporally non-malleable, then EUT straightforwardly excludes planning difficulties. There is no scope to plan, let alone encounter planning difficulties (a sub-class of negative material secondary satisfactions) if the axiomatic justification of EUT applies only to a world in which *nothing* is inter-temporally transferable, the case if a non-inter-temporal transferability restriction has to be imposed on its outcomes space!

In summary, regardless of whether we examine the first or the second version of the proposal to redefine the outcomes space as concerning more "primitive" preferences, the conclusion is the same. The redefinition deprives EUT of axioms that are plausible, that include material secondary satisfactions. The almost universally shared view is that the proposed "primitive" redefinition solves the problem theoretically, but not practically, since it is too cumbersome to go down to the primitive level. The above reveals that this view of the "primitive" redefinition proposal is, on examination, false. Going down to a "primitive" outcomes space, even were it feasible, does nothing to retrieve for EUT an axiomatic base - a set of plausible axioms implying equation (3.5) and the concomitant restrictions, where plausible here means taking into account planning costs under risk and other instances of material secondary satisfactions.

6.2 Temporal EUT

The planning problem, the failure of EUT in the temporal context, was long ago recognized. There have also been numerous efforts to avoid EUT failing, with others noting explicitly or implicitly where these efforts fell short of EUT furnishing sensible answers in the case of planning difficulties. Contributions include Wold (1952), Markowitz (1959), Dreze & Modigliani (1966, 1972), Mossin (1969), Eden (1977), Munera (1978), Rossman & Selden (1979), Kreps & Porteus (1978, 1979a,b), De Neufville & McCord (1983b), De Neufville & McCord (1983a), Pope (1983), De Neufville & Delquie (1988), De Neufville & Odoni (2002). Keynes' objections to Ramsey's embryonic EUT and its non-axiomatised forebears also relate to planning problems and uncertainty effects yielding negative secondary satisfactions (Walsh, 1996, pp 56, 52-65).

The special contribution of Kreps & Porteus (1978) is to be one of a small group who were time-wise careful about the EUT axioms. They belong to the small group who recognise that atemporal EUT axioms cannot address planning problems. They extended the axioms of EUT to distributions over outcomes with dated onset and expiry dates of non-degenerated probabilities. It is invaluable and pathbreaking work to investigate, as Kreps and Porteus did, how far EUT might go when temporal axioms are devised to yield the expected utility property in relations that date the onset and expiry of each probability.

But in so extending EUT they had to introduce new axioms that, as they themselves note, will not in general be plausible or easy to check whether they hold. Further these Kreps and Porteus set of extended EUT axioms preclude many instances of secondary satisfactions. For instance their set of extended EUT axioms requires a globally transitive ranking and a between-ness substitution axiom, respectively axioms 2.1, 4.1 and 2.3, 4.3 in Kreps & Porteus (1978). But thresholds in loan default rates for issuing loans introduce intransitivities into material secondary satisfactions and preclude this between-ness substitution axiom being true. If you cannot guarantee to repay, another may refuse to lend you money for dinner so that you miss out on this material secondary satisfaction. If you will miss out on the loan anyway, you may well prefer the very risky alternative to a fairly risky alternative. This will often be the case even though you prefer the safe to the very risky. In such cases you violate both these axioms. In addition, other forms of secondary satisfactions besides those from risk-based inter-temporal consumption transfers have to be independent of these transfers. This is rarely the case: thrills, worry, blame, ridicule are normally related to the consumption / asset outcomes flow, virtually never independent of it.[2]

Kreps and Porteus concluded that such an extension of EUT might be an acceptable (albeit imperfect) model of induced preference in a wide variety of applications (Kreps & Porteus, 1979b, p 101). But this conclusion runs counter to their own observations on the peculiar utility function to which it would restrict analysis. It also runs counter to their own observations on the limited plausibility of their axioms for the reasons to which they drew attention. Further it runs counter to a wider look at secondary satisfactions, and the problem that large classes (indeed arguably most of them) are excluded under these actions.

[2]There is in addition a quite sizable literature that deals with borrower-lender issues related to changing knowledge - under the assumption that this could be done within EUT, eg Stiglitz & Weiss (1981), Gale & Hellwig (1985), Kiyotaki & Moore (1997). This literature has extremely important results - but ones outside the axiomatic base, for the reasons shown in eg Kreps & Porteus (1979b) and Pope (2005b).

6.3 The Missed Connection of Temporal EUT With the von Neumann-Morgenstern Complementarities

Researchers investigating the planning problem did not notice that the sub-optimal consumption-investment decisions under risk are, in the language of von Neumann and Morgenstern, a negative complementarity. The good and bad outcomes interact to cause the chooser to ex post have consumed too little (on learning in the post-outcome period that the good outcome has occurred), or too much (on learning that the bad outcome has occurred). Thereby researchers investigating the planning problem did not connect the planning problem with the von Neumann and Morgenstern contradiction. Nor did they recognize that they were *not* dealing with a *special* subset of risky acts - ones which they termed temporal lotteries and contrasted with atemporal lotteries. Rather they were dealing with the *only* sort of risky choices that exist, temporal ones.

Without a *pre-outcome* period, there is no risk, the outcome is sure. All lotteries are temporal. Thinking of the planning problem as unusual in involving temporal lotteries also limited the non-material instances investigated and discussed. There was non-recognition of other material satisfactions based on *degree of knowledge ahead* such as social security eligibility, and ability to commit and obtain loans, jobs (Pope (1983, 1991b, 2004, 2005b)).

Chapter 7

Misconceptions About Secondary Satisfactions

There have been six objections to including consideration of secondary satisfactions in contract analyses.

7.1 Inappropriate Tests of Secondary Satisfactions

The first objection is that the regret and rejoicing model of Loomes & Sugden (1982) is not the best predictor of experimental evidence. This is no argument against secondary satisfactions playing a crucial role in choice. It is rather evidence against Loomes and Sugden's: a) functional form for two secondary satisfactions, namely regret and rejoicing, b) exclusion of all the other secondary satisfactions that would normally confound identification of regret and rejoicing effects, and c) assumption that regret and rejoicing are context and scale independent. It should also be noted that the designs employed an inappropriate set-up for comparing EUT's relative predictive power. As noted in chapter 2, with inappropriate cage set-ups of Skinnerian behaviourism, Darwin's hypotheses on the importance of secondary satisfactions could not be investigated. Likewise sufficient time is essential in the laboratory set-up for detecting related effects, *eg* Hassenzahl & Borcherding (2004).

7.2 Misleadingly Denigratory Examples of Secondary Satisfactions

A second objection to including secondary satisfactions is that they are misperceived as exclusively pathological or frivolous or trivial or absent from serious choices, or even unethical, eg Ramsey (1926, 1950), Marschak (1950). This misperception is partly a function of the social gambling contexts many scientists used for understanding them. Once more serious contexts are used, more serious examples of secondary satisfactions come to light. These include secondary satisfactions like hope and fear that many religious traditions deem must as a matter of ethics be taken into account, Pope (1983). Good decisionmaking requires emotional inputs from secondary satisfactions like fear (Damasio (1984)) and regret (Camille et al. (2004)).

Once more serious examples are used it also becomes apparent that quantities matter, and that we need to distinguish between descriptive choice theory and prescriptive. For descriptive choice theory, we ought consider what anticipated satisfactions matter to the person and drive his choices, not what anticipated satisfactions we deem ought drive his choices. For prescriptive theory, the ought enters.

A person unable to worry enough or feel enough regret may well be a pathologically bad decision maker because too foolhardy, as Damasio (1984) and Camille et al. (2004) found. There are also persons too worried or too regretful or too hopeless to make good decisions. Indeed there are pathologically low and pathologically high levels of many, perhaps all, emotional secondary satisfactions.

In the same way there are pathologically low and high levels of primary satisfactions. The person can have a pathologically low level of primary satisfactions from eating (the anorexic) or a pathologically high level of primary satisfactions from eating (the obese). Again, a person can have a pathologically low level of primary satisfactions from water (the dehydrated) or a pathologically high level of primary satisfactions from alcohol (the drunk).

In short, people can have pathological levels of some primary satisfactions. But we do not conclude that *all* primary satisfactions are to be ignored, excluded from prescriptive choice theory. For prescriptive choice theory, we need to take the same attitude to secondary satisfactions. We need to discriminate between levels that are normal, even desirable, and levels that are pathological/irrational (according to our benchmark).

We should also bear in mind the observations of Bell (1981), that in prescriptive theorizing, anticipated irrational reactions that cannot be overcome, ought influence choice. Suppose a person knows that he cannot restrain himself from becoming pathologically worried if he chooses a risky act and that this will decrease his wellbeing in the pre-outcome period. Then it may well be better for him to avoid that risky act - even if he knows that a more rational / less pathological being would not suffer any significant amount of the negative secondary satisfaction of worry in the pre-outcome period.

As discussed in chapter 6, there are also material and financial secondary satisfactions such as access to pensions, Pope (1983), and effects of ability to commit to repay a loan, Pope (1991b) and to commit to a job, Pope (2004). No economist would describe these secondary satisfactions as trivial. Their neglect is primarily that with a model of the world like that of Figure 5.1 inspired by EUT, theorists simply lacked a *pre-outcome* period in which to notice that there are major material secondary satisfactions that cannot even be defined, let alone modelled without incorporating *knowledge ahead.* There is for instance no such thing as a loan, unless there is a commitment to repay. Commitment is about the future, a *knowledge-ahead* phenomenon, whose modeling in game theory violates its axiomatic justification, namely EUT, Pope (2004, 2005b).

7.3 Multiple Choice Fallacies

An objection to modelling financial and material secondary satisfactions is to allege that the modelling involves multiple acts, whereas EUT was deemed in 1952 to require that choosers only make one decision, Savage (1952a) and Savage (1952b). This line of reasoning became ensconced by Savage (1954) with the assertion that it is irrational not to "look before you leap" and make more than one decision in the whole of life. You should make a single grand decision, amalgamating into this grand decision all simultaneous and all sequential decisions you might make at different times throughout your conceivable life. After this, you never make another decision or you would be irrational - someone who had not properly looked before you leapt!

This grand single decision is impractical as even Savage conceded.[1] But to avoid distraction, the examples cited above in Pope (1983, 1991b, 2004) were explicitly and artificially such grand single decisions. But these were theoretical papers. For a book reporting laboratory work, it would be too patently contrary to the facts and against the experimental method to invoke the single grand decision. To invoke it is to propose that participants imagine and attempt to perform choices in our laboratories as if one out of their multiple hypothetical laboratory choices will be the *only* choice in their whole future life.

Accordingly, we did not state that instead of making an insurance decision and a decision on a loan and dinner out, these are combined with decisions on how much to eat for breakfast tomorrow, ... whether to catch the train or bicycle back from the office a week hence, ... how many jars of coffee to buy in 2010, how many stocks to sell in the year 2015 if the price drops to z, and ... , in the form of a grand single decision for the whole of life. Note however, that the problem could be so formulated, and thus that the objection to multiple choices is never a valid objection against including material and financial secondary satisfactions in the decision procedure. The objection is as groundless as saying that a rational choice theory cannot apply to whether to buy bread today since it is to be applied to buying butter today. The problem can be formulated on whether to buy bread only, butter only or bread and butter. It can equivalently be formulated as whether to buy bread today given that in the shop a decision will also be made at the same time or an irrelevantly later time after, on whether to buy butter. Rational choice requires the chooser to take into account foreseeable interconnected decisions. The butter (and its price etc) can be rationally taken into account in the bread decision, no matter whether the problem is formulated as two distinct decisions or as a joint bread and butter decision. So it is also

[1]Savage sought a way out of the difficulty, pointed out to him by Wold (1952). See Savage (1952a, 1952b) and Samuelson (1952b). In 1954, it still remained elusive. Savage proposed a "small worlds" assumption for cases not involving the intertemporal interdependencies of secondary satisfactions, but even for these cases he could not work out the conditions under which it would be valid but left this task to future researchers.

with interconnected decisions such as insurance, and abilities to commit and intertemporally plan.

7.4 Inconsistent Elaborations of EUT

A fourth objection to extending EUT to include emotional, material and financial secondary satisfactions is the claim that there is no need to extend it, that secondary satisfactions are already in EUT. There are several different phrases used for saying how secondary satisfactions are already included in EUT. One is to say that they are in EUT once the outcomes (and hence acts) are properly elaborated / defined, *eg* Samuelson (1952a), Markowitz (1959), Caplin & Leahy (2001). Another is to say that they are in EUT once the decision situation is fully described including all its opportunity costs, *eg* Luce & Raiffa (1957), Stiglitz & Weiss (1981), Gale & Hellwig (1985), Kiyotaki & Moore (1997). The third is to pronounce that all satisfactions that should be in are already included under EUT's "consistency requirements", Savage (1954)[2], or even by "definition", *eg* Ellsberg (1954) quoted approvingly in Schoemaker (1982). Among scientists holding this view, some state explicitly that von Neumann and Morgenstern erred in their interpretation of EUT, *eg* Ellsberg, and some express puzzlement and disappointment that von Neumann and Morgenstern disagree with them, *eg* Savage (1954). Most younger scientists have never heard about the heated disputes of the early 1950s on whether EUT omits anything that matters, and thus have no notion that their interpretation of EUT is the reverse of that held by von Neumann and Morgenstern.

But as Bell (1981) and Loomes & Sugden (1982) implicitly assert, and as explicitly addressed and proved in Pope (1983, 1984, 2000, 2004, 2005b), von Neumann and Morgenstern had the correct interpretation, namely that EUT excludes secondary satisfactions. Including secondary satisfactions in EUT destroys its axiomatic base. Secondary satisfactions cannot be consistently included in EUT's utility mapping. Including secondary satisfactions robs EUT of the scope to identify a unique act (or unique set of best acts) in each given choice set.

[2]Savage had by then radically altered his interpretation of EUT from that in Friedman & Savage (1948). This earlier piece states that EUT omits secondary satisfactions and that this omission might be an empirical weakness.

7.5 Violations of the Dominance Principle

The fifth objection to secondary satisfactions is that they can lead to violations of the dominance principle. The dominance principle is obeyed by EUT and what we here term EUT+ theories, a much wider set of theories. EUT+ includes cumulative prospect theory, Tversky & Kahneman (1992), Bleichrodt et al. (2001), the anticipated utility theory of Quiggin (1982, 1993), the invariance theory of Allais (1988) and indeed most rank dependent generalizations of EUT. EUT+ even includes even some theories seeking to include particular secondary satisfactions, eg the disappointment theory of Gul (1991) and the intrinsic preference for information theory of Grant et al. (1998).

It is admitted informally by some that imposing the dominance principle is unnaturally and implausibly confining when modeling secondary satisfactions. Why then are there so few scientists interested in secondary satisfactions ready to drop the dominance principle? Why is it that proposals recur, such as the elaborated outcomes of Markowitz (1959), for trying to avoid violations of the dominance principle and thus of EUT, while admitting the importance of secondary satisfactions in their own decision making - in Markowitz's case his choice of acts that allow him the pleasures of wondering and surprise?

The answer is that the dominance principle is almost universally deemed to be (i) rational and (ii) something people never wish to violate in transparent situations, such as simple two act choice sets. This creates cognitive dissonance for many scientists who recognize and wish to model secondary satisfactions. For the few for whom there is no such cognitive dissonance, there is another hurdle - getting their new models accepted if they violate the dominance principle while it stands on such a normative and descriptive pedestal.

What then is the esteemed dominance principle? The dominance principle states that people should always prefer acts with the property of FOSD. In turn FOSD is a a first order stochastically dominating distribution of outcomes. Outcomes are here defined and must be defined *independent of knowledge ahead*, ie independent of the probabilities of the outcomes. Otherwise we cannot compare two (de)cumulative distributions and determine which dominates.

To illustrate FOSD, consider two outcomes. One outcome is nice, namely seeing good scenery and returning home. The other outcome is horrid, seeing less good scenery and not return home, instead dying on the mountain in an avalanche. Consider now choice between two climbs. One climb is on Mount Pleasant that is a sure act. It guarantees the nice outcome of seeing good scenery and returning home. The other climb is on Mount Danger. It only almost guarantees the nice outcome. With it there is a whiff of danger, a tiny risk of the horrid outcome with death. Then Mount Pleasant has the FOSD property. Choosing it one cannot get an inferior outcome and may get a better outcome than choosing Mount Danger. Choosing it obeys the dominance principle.

But is it wise for people and societies to invariably obey the dominance principle? Has every mountaineer made an error every single time she disobeyed it? Happiness and brain health requires brain activity - avoiding boredom by taking appropriate risks. Aspects of this need for stimulation from risks (ie from suspense, from wondering and then being surprised, from challenges), have been identified by Pascal (1670), Darwin (1874), Hebb (1949), Samuelson (1952a), Markowitz (1959), Scitovsky (1976). Everyone needs thrills, and needs to find socially and individually appropriate ways of getting their needed amount of thrills. Otherwise their brain become dull, ill.

It may well be desirable to encourage young lads into mountaineering and away from car theft as a means of getting these thrills. Ie disobeying the dominance principle and going up Mount Danger may be for such youngsters the appropriate choice. Pope (2000) questioned faculty in economics and the decision sciences and found that in such a situation concerning serious matters of brain health and car theft, most would advise disobeying the dominance principle.

See Pope (1991b, 2005b) for financial secondary satisfactions pertaining to loans generating rational violations of the dominance principle. See Bernard (1966) for perhaps the first articulation of why the dominance principle should be dropped in both prescriptive and descriptive theorizing on risky choice. In short, the dominance principle only seems sensible, something to be invariably obeyed, when the scenario is left excessively abstract. Then people lack the imagination to discern when *knowledge-ahead-based* sources of satisfactions like positive ones from thrills, exploring, and negative ones like fear or loan ineligibility, are ignored.

7.6 Introduction of Contradictions

The final objection to including secondary satisfactions is that secondary satisfactions introduce a contradiction. This is the contradiction that von Neumann and Morgenstern reported finding in their efforts to include them. (See Chapter 5 above.) The allegation is that secondary satisfactions are akin to "mistakes of arithmetic", *eg* Marschak (1950).

There are three ways of tackling this objection. One is to tackle it from first principles. Argue for the reasonableness of including consideration of at least some of them in prescriptive and normative work. This way is to be found in *eg* Bell (1981), Loomes & Sugden (1982), and Pope (1983, 1984).

The second is to argue against the notion that every primary satisfaction is rational, and every secondary satisfaction irrational. This can be done by providing instances in which a normative theory might wish to deem irrational some primary satisfactions like urges to overeat, over drink and such like, and deem rational some secondary satisfactions like thankfulness, and ability to commit, *eg* Pope (1991b, 1995). In short these are demonstrations that there has been an irrational one-sided focus on the lure of compulsive gambling and other secondary satisfactions that most of us would deem irrational, and a failure to focus on the many primary satisfactions that most of us similarly deem irrational.

The third way of tackling the contradictions claim is to recognize, like von Neumann and Morgenstern, that this as a pseudo contradiction, awaiting the higher level of analysis that would dissolve it. As shown in Chapter 5, that higher level is the evolving stages of knowledge ahead, a distinguishing feature of the act being risky. In short, SKAT, the Stages of Knowledge Ahead Theory dissolves the contradiction. The contradiction arises only from the imperfect transfer of choice under certainty into choice under risk of traditional theories - an imperfect transfer that excluded the fundamental feature of risk, an anticipated change in knowledge ahead.

Chapter 8

Biases in Experimental Set-ups That Ignore Secondary Satisfactions

As the great Australian biologist Anthony Barnett observed, the experimental set-up largely dictates the results observed and the conclusions drawn. Eventually he and his co-researchers were able to overcome scientific opposition and obtain national science foundation funding to go beyond Skinnerian behaviourist cage set-ups for their rats, and investigate Charles Darwin's hypotheses on animals' positive secondary satisfactions, *ie* satisfactions derived from discerning that they have only a limited degree of *knowledge ahead.* He discerned the need of animals for excitement, their love of exploration and curiosity. Such experiences are excluded if the world seems certain, in which case as Darwin (1874) noted, the animals suffer from boredom. Barnett's set of studies, and related ones by French scientists on cockroaches confirmed Darwin, and discovered also the importance of negative satisfactions from having only a limited degree of *knowledge ahead*, namely fear from too much risk. See *eg* the retrospective in Barnett & Cowan (1976).

By and large experimental economics has the EUT Skinnerian touch, viewing its participants and their decisions through theoretical lenses bereft of knowledge-ahead-*based* effects, whether material or emotional. Under the EUT Skinnerian knowledge ahead-*independent* framework it does not matter how long is *pre-outcome* period, *ie* how long the choosers anticipate waiting after choosing the risky act of no insurance making a decision for or against taking out insurance and learning whether the bad outcome occurs.

Most experimental analyses of choice under risk are conducted as if the duration of this risky *pre-outcome* period is irrelevant. They are conducted as if it does not matter whether the choice of a risky act imposes on the contracting parties an anticipation of limited *knowledge ahead* for a decade, 24 hours, two hours, or a minute. When experimenters analyse and construct models largely on the basis of hypothetical situations (*eg* prospect theory), participants lack information on the duration of the *pre-outcome* period of *limited knowledge ahead.*

Should participants answer anticipating that any risk will be almost instantly resolved? Or should participants anticipate that as in so many real life choices, they would wait years or even decades before learning whether their investments were profitable? That is, should participants answer for the case where k, the duration of the *pre-outcome* period of *limited knowledge ahead* is ultra short or exceedingly long? At a minimum, this ambiguity about the duration of k generates unnecessary noise in the answers elicited, and unwarranted claims on the universality of the findings to real world situations with differentially lengthed *pre-outcome* periods.

In experiments actually played out in the laboratory, there is a reduction in such unnecessary noise from differently lengthed *pre-outcome* periods entering choices in analyses and models constructed from data in which participants (probabilistically) receive what they choose. Here participants typically have a reasonable idea of the *pre-outcome* period length, *ie* of when they will learn the outcome of their choice normally within an hour or so at the end of the session. But since the experimenters still ignore how long the *pre-outcome* period is, they inadvertently unwarrantedly infer that the participant's choice with a pre-outcome period of the short duration of roughly an hour (in the average experimental setup) is the same as in other situations in which the pre-outcome period can have very lengthy durations.

The problem is much deeper than one of ignoring how differentially lengthed pre-outcome periods alter decisions. The problem is that scientists, who ignore the impact of the length of the pre-outcome period, analyse and construct risky choice models as depicted in Figure 5.1. That is, they analyse as if the *pre-outcome* period did not exist at all - as if people who decide not to insure could jump with a zero time lapse from when she chooses at $t = 0$ into the last day of their future. Experimenters analyse decisions ignoring the set of material, financial and emotional sources of secondary satisfactions.

Failure to recognise the very existence of the *pre-outcome* period, and thus to give participants leads on its likely cause-effect ramifications, hampers scientists in eliciting meaningful answers from participants and distilling from them useful descriptive models. This causes the following problems.

First, scientists are unable to recognise that their measurements are contaminated by the omitted variables bias of having overlooked the entire class of secondary satisfactions. *Ie* they have overlooked all cause effect chains that participants anticipate from going through a *pre-outcome* period of limited *knowledge ahead.* The adverse consequences of overlooking these cause-effect chains include the following:

a) Their measures of the utility of primary satisfactions exhibit systematic pronounced biases. To take one example, consider experimental efforts at eliciting the shape of a participants' EUT curve of the utility from outcomes independent of *knowledge ahead.* If the experimenter uses the expected utility theory standard gamble elicitation procedures, these omitted variables effects give the curve obtained an inflection from concave to convex under the probability equivalence method of Ramsey, and the reverse twist under the certainty equivalent method of von Neumann and Morgenstern, whereas under the assumptions of these two measures, the curves should be identical, Pope (2004).

b) Scientists engage in counterproductive experimental set-ups because they do not understand the sources of contamination in their data. To take one example, scientists have wished to analyse ignoring the concavity or convexity of participants' EUT curve of the utility from outcomes independent of *knowledge ahead.* They have assumed that they can analyse ignoring it if the lotteries only have two outcomes. After all, three outcomes are needed to detect concavity and convexity. But because there are unrecognised *knowledge-ahead-based* effects contaminating the data, and these effects as measured by variance become more pronounced under binary lotteries (constructed *eg* with the same range and expected value but no middle outcome), binary lotteries do even worse as regards the experimenters' goals, Abbink et al. (1999).

The second problem that arises from failing to recognize the existence of the pre-outcome period is in the analysis made of participants. When they do not recognize the pre-outcome period, experimenters misconstrue as inconsistencies, irrationalities, contradictions, many choices made by participants. Some are reasonably taking some account of the anticipated effects on them of going through this risky *pre-outcome* period. Experimenters cannot distinguish between these and cases of genuine irrationality. To take one example, participants' responses to having limited *knowledge ahead* in the *pre-outcome* period cause them to rationally choose (stochastically) dominated acts in some choice sets, Bernard (1966, 1970), Pope (1984, 2001). The experimenter sees this as irrational behaviour. But once *knowledge-ahead-based* satisfactions are taken into account, it can be shown that EUT is irrational in that it forces the chooser to take an act with a lower expected level of satisfaction in *some* choice sets. Instances are in Pope (1991b, 2004).

Experimenters are good empiricists. They discern motivations missing in standard theory. But without a *knowledge-ahead-based* conceptual framework with a division of the period following choice into a *pre-* and *post-outcome* period, they will miss motivations, and lack an adequate means of identifying choices that really are irrational and that generate genuine contradictions as distinct from *pseudo* contradictions. They will deem that participants in their experiments are contradicting themselves, when such contradictions are *pseudo*, due to the limitations of analysts' getting their own conceptual framework on the right level.

Chapter 9

SKAT, The Stages of Knowledge Ahead Theory

We wish to analyse risky choice unbiasedly and consistently and with sufficient breadth to allow all broad classes of secondary as well as primary satisfactions. To do so we need to go not merely beyond static EUT and its atemporal generalizations such as standard rank dependent theory. We also need to go beyond those models designed to approximate aspects of secondary satisfactions without delineating changing stages of knowledge ahead as is the case eg with most regret theories. We need information of when risks are resolved, but not, as in eg Kreps and Porteus, associated axiomatic constraints. As noted in Chapter 6, such constraints irrationally exclude many secondary satisfactions, and relegate the role of other secondary satisfactions to being, in effect, the accidental outcome of a different maximization procedure - one that *accidentally* approximates for *some* planning situations for *some* individuals.

We require a decision theory that gives balanced weight and appropriate prominence to secondary primary satisfactions, and thus does not merely consider the subset of primary satisfactions reaped after risk is in the passed. SKAT, the Stages of Knowledge Ahead Theory gives this rational balance. This is presented briefly in Pope (1983) and in more detail in Pope (1995, 1996/7). Its distinctive feature is its partitioning of the future by changing stages of knowledge ahead as discussed already in Chapter 5 in connection with the resolution of the von Neumann and Morgenstern contradiction.

In minimal form for the simplest risky choice scenarios, such as that of the experiments analysed in this book, this epistemic partitioning involves only one pre-outcome period (before he will know the outcome of his chosen act), and only one post-outcome period (the more distant part of his future) when he will have learned the outcome of his chosen act as in Table 9.1:

Table 9.1: The Pre-Post Outcome Distinction

Pre-outcome period	
Zero duration	if chooses insurance since upon choice at $t = 0$ knows the outcome segment after X hours, \$85
X hours duration	if chooses to stay uninsured, risking loss of \$100 if there is an attack
Post-outcome period	
Begins upon choice	if chooses insurance, since upon choice at $t = 0$ knows the outcome segment after X hours, \$85
Begins after X hours	if chooses to stay uninsured, and is \$0 if there is an attack, otherwise \$100

This epistemic partitioning allows the SKAT chooser to identify two sorts of causal chains, namely:

(i) those with anticipated effects caused by each current segment of the outcome flow, namely primary satisfactions, and

(ii) those with anticipated effects caused also by what seems possible or used to seem possible, namely secondary satisfactions.

An illustrative list of secondary satisfactions anticipated in the pre- and post-outcome periods is in Table 5.2 of Chapter 5 for the case of the risky act of deciding to stay uninsured. The SKAT chooser augments this list with a second list. This second list is of his anticipated primary satisfactions from his current situation alone in both period 1 and period 2. Thereby the SKAT decision maker takes into account both the major primary satisfactions $Ppre_h$, $h = 1, 2, \ldots$ and the major secondary satisfactions $Spre_j, j = 1, 2, \ldots$ that he anticipates will *simultaneously* ensue in the *pre*-outcome period.

Note the "ands" of the pre-outcome period anticipated satisfactions in Table 5.1, denoting this simultaneity. The SKAT decision maker also considers the major primary satisfactions $Ppost_{mi}, m = 1, 2, \ldots$ and secondary satisfactions $Spost_{ni}, n = 1, 2, \ldots$ that he anticipates will ensue in the *post*-outcome period. Some of these he anticipates will be *simultaneously* reaped, all those that hold for a given outcome i. Some pertain to different outcomes i, and thus are *mutually exclusive.* Note the mix of "ands" and "ors" of the *post*-outcome period anticipated satisfactions in Table 5.2.

The SKAT decision maker then requires atemporal aggregation rules for amalgamating all these satisfactions that he has listed into an overall value. He needs the overall value either to assess whether that act is good enough (if he is a satisficer), or to compare with those alternative acts in the choice set that he considers warrant analysis. The SKAT decision maker's atemporal aggregation / amalgamation can be divided into the following four steps:

Step 1 in the SKAT aggregation:
$Vpre$ Amalgamation into some value index of the different satisfactions that are anticipated to be experienced *simultaneously* in the *pre*-outcome period. Some of these will be primary satisfactions $Ppre_h, h = 1, 2, \ldots$ and some of them secondary satisfactions $Spre_j, j = 1, 2, \ldots$.

Eg if each such satisfaction mapped into a number, and if primary and secondary satisfactions are additively separable, this could be their weighted sum, with weights respectively w_h and w_j.

$$SKAT : Vpre = \sum_h w_h Ppre_h + \sum_j w_j Spre_j. \tag{9.1}$$

w_h	atemporal aggregation weight for $Ppre_h$, *outside time*
$Ppre_h$	anticipated primary satisfaction in the pre-outcome period, in *real time*
w_j	atemporal aggregation weight for $Spre_j$, *outside time*
$Spre_j$	anticipated secondary satisfaction in the pre-outcome period, in *real time.*

Step 2 in the SKAT aggregation:
$Vpost_i$ Amalgamation into some value index of the different satisfactions that are anticipated to be experienced *simultaneously* in the *post*-outcome period under each mutually exclusive outcome $i, i = 1, 2, \ldots$. Some of these will be primary satisfactions $Ppost_{mi}, m = 1, 2, \ldots$ and some of them secondary satisfactions $Spost_{ni}, n = 1, 2, \ldots$.

Eg if each mapped into a number, and if primary and secondary satisfactions are additively separable, this could be their weighted sum, with weights respectively w_m and w_n,

$$SKAT : Vpost_i = \sum_m w_m Ppre_{mi} + \sum_n w_n Spost_{ni}. \tag{9.2}$$

w_m	atemporal aggregation weight for $Ppost_{mi}$, *outside time*
$Ppost_{mi}$	anticipated primary satisfaction in the post-outcome period, in *real time*
w_n	atemporal aggregation weight for $Spost_{ni}$, *outside time*
$Spost_{ni}$	anticipated secondary satisfaction in the post-outcome period, in *real time.*

Step 3 in the SKAT aggregation:
Vpost aggregation into some value index of the *mutually exclusive* primary and secondary satisfactions anticipated in the post-outcome period $Vpost_i$.

Eg if each index mapped into a number to which the chooser attaches a numerical probability, $p_i, i = 1, 2, \ldots$, this could be their probability weighted sum,

$$SKAT : Vpost = \sum_i p_i Vpost_i \tag{9.3}$$

p_i	atemporal aggregation weight for $Vpost_i$, *outside time*
$Vpost_i$	anticipated summed primary and secondary satisfactions, in the post-outcome period for outcome i, in *real time.*

Step 4 in the SKAT aggregation:
V aggregation into some value index of $Vpre$ and $Vpost$.

Eg if each mapped into a number, and if these are temporally linearly separable, this could be their weighted sum with weights respectively w_{pre}, and w_{post},

$$SKAT : V = w_{pre} V_{pre} + w_{post} V_{post} \tag{9.4}$$

w_{pre}	atemporal aggregation weight for V_{pre}, *outside time*
V_{pre}	anticipated summed primary and secondary satisfactions, in the pre-outcome period, in *real time*
w_{post}	atemporal aggregation weight for V_{post}, *outside time*
V_{post}	anticipated summed primary and secondary satisfactions, in the pre-outcome period, in *real time.*

These four steps can be quantitative as in the examples given above for each and for the resultant overall value of the act V. When EUT is stated in the terminology of primary and secondary satisfactions, each with multiple components, it limits consideration to the first weighted sum in equation (9.2) entering equation (9.3), ie,

$$EUT : V = Vpost_i = \sum_i p_i(\sum_m w_m Post_{mi}) \tag{9.5}$$

p_i	atemporal aggregation weights for $Ppost_{mi}$, *outside time*
$Ppost_{mi}$	anticipated primary satisfaction in the post-outcome period, in *real time.*

The formulation of SKAT in equation (9.4) is used above to highlight the distinctions from EUT in equation (9.5) *even in the case of numerical satisfactions and numerical weights and linearly separable satisfactions and numerical probability weights for mutually exclusive satisfactions.* More plausibly in employing SKAT in virtually all problems, including the insurance scenario investigated in this book (see Chapter 14), steps 1 to 4 are conducted qualitatively.[1] This is because often the aggregation, like the satisfactions themselves, will be of a qualitative form, and may follow algorithmic procedures of sequentially considering sources of satisfactions, often with a satisficing component, choosing an act when it yields enough satisfaction. The general form of SKAT is equation (9.6),

$$SKAT : V = V(w_h, Ppre_h, h = 1, 2, \ldots, w_j Spre_j, j = 1, 2, \ldots, w_i,\\ i = 1, 2, \ldots, w_m, Ppost_{mi}, m = 1, 2, \ldots, w_n, Spost_{ni}, n = 1, 2, \ldots) \tag{9.6}$$

This allows for interdependencies of different satisfactions and also for the norm of non-quantitative satisfactions and weights, ie allows also :

(i) *qualitative* pre-outcome primary and secondary satisfactions and associated weights, $Ppre_h$, $Spre_j$, w_h and w_j,

(ii) *qualitative* post-outcome primary and secondary satisfactions and associated weights, $Ppost_{mi}$, $Spost_{ni}$, w_m, w_n and w_i,

(iii) *qualitative* weights for aggregating the pre-and post-outcome satisfactions, w_{pre}, and w_{post}.

In many situations only a few of the primary and secondary satisfactions are major enough or anticipatable enough to warrant attention and get any sort of weight. What matters for reasonable choice in most situations is not how quantitative is the information used by a decision maker.

[1] Further as Simon (1955) proposes, of which an application is Sauermann & Selten (1962), Selten (1998), there will normally be limited investigation of what enters the choice set, and even of acts already identified as within the choice set. It will be frequently the case that the sensible time to stop analyzing alternatives is when a satisficing act is located.

What matters is that he has not overlooked a major source of positive or negative satisfactions. Atemporal frames fail to pose the chronology of an anticipated change in knowledge ahead which occurs for all risky acts. Atemporal frames and inconsistent notions that there can be atemporal lotteries, render it difficult for choosers to notice whether they have overlooked a major source of positive or negative satisfactions. Laying out the secondary satisfactions by *changes in the stage of knowledge ahead* as illustrated in Table 5.2, with an extra column for the primary satisfactions, reduces this problem of overlooking important secondary satisfactions. It is moreover hard when secondary satisfactions all stem from which particular *stage of knowledge ahead* the chooser will be in, to notice secondary satisfactions without laying out the future formally or informally in this epistemic (stages of knowledge ahead) manner.

The exactness of these anticipations of primary and secondary satisfactions, and their nature will then largely determine the sorts of ways in which it can be appropriate for the chooser to do the atemporal aggregations. Different aggregation procedures related in part to different degrees of interdependence between the different satisfactions, constitute different models within SKAT, the Stages of Knowledge Ahead Theory. Employing SKAT aids scientists in classifying when choosers seem foolish, when rational. Without it, scientists can fall into the trap of not recognizing when choices are sensibly guided by secondary satisfactions.

As we saw in Chapter 8, without SKAT for consistently discerning secondary satisfactions, scientists can fall into the trap of thinking that a preference of stochastically dominating distributions over outcomes is rational and that to do otherwise foolish. To the contrary, the sensible decision maker taking into account issues of worry, thrills, blame, ridicule, inability to commit, planning problems will violate this dominance principle in certain choice sets. Ie, the dominance principle is a principle that only primary satisfactions matter. It seems such a sensible principle when we fail to notice the existence of the pre-outcome period both in how we frame problems theoretically and describe decision situations to our participants. No secondary satisfactions are mentioned. Only the primary satisfactions, the payoffs, with typically no mention of when the payoff is to be reaped making it readily seem that no secondary satisfactions could exist.

Chapter 10

Self Reports *Versus* Scientists' Own Introspection and Related Reliability Issues

We shall base our assessment on choosers' decision procedures importantly on their self-reports of their grounds for choice. Self reports tend to be denigrated by economists. Below we survey the various grounds for objection, and show that from a scientific viewpoint, the objections are flawed. Objectors often offer a variant of Alchian's thought competitiveness experiment (Alchian (1950)) and consider that it does not matter how the participant decided, that only those who use EUT survive, with the rest eliminated by competition. The argument is a derivative of the criticism of Cyert and March's discovery that firms do not maximise profits (Cyert & March (1963)). The proposal that there are no non-profit maximising firms since competition eliminates all who do not accidentally maximise profits has long since been shown to be false, even in a hypothetical competitive world, once risk and uncertainty are *systematically* modelled. Indeed once risk and uncertainty are systematically modelled, the issue of identifying temporally what is the expected profit flow to maximise has yet to be satisfactorily solved. Similar comments apply to evolutionary models seeking to extend this assertion to EUT as distinct from expected profits. As demonstrated in Pope (1991b) with respect to loans, and with respect to jobs (Pope (2004)), using EUT can lead to minimising not to maximising welfare and income, and thus cannot be presumed to constitute the evolutionarily surviving strategy, not even in the short run.

Another common objection to using information from participants' reports on why they chose is that these reports are fallible. Objectors may point to psychological and neuronal experiments proving that some decisions are made before choosers realise that they have decided and giving implausible if not patently false accounts of why they chose. All this is true. Reported reasons are not infallible. However the classic examples cited in this regard are not too pertinent to the demand for protection contracts. Nor indeed pertinent to most choices assumed to be made under EUT and its generalisations.

Classic examples proffered of the participant's reasons for choice being unreliable stem from four factors:

1 There has been some considerable time lapse since they chose,

2 what people are asked to recall is complicated,

3 people wish to conceal their actual motives,

4 the decision was unconscious and so what determined it cannot be unearthed by the conscious mind.

The first three factors will not be salient in our situation, except in so far as they seek to report having used the principal methods taught at universities, namely EUT or its simplification, expected value. Insofar as this teaching entices them to report usage of EUT when participants do not actually use it, the answers are biased toward (not away from) the conclusion that EUT is relevant to the demand for contracts.

Consider now the claim that no participant knows anything about how he chooses since this is all unconscious. There may indeed be unconscious elements in the procedure. But then the question arises how has any choice theory been given any credibility in the decision sciences if we reject self-reports as useless since so unreliable.

The actual situation is that those who repudiate self-reports believe that some scientists (in particular those who advocate EUT) know how decisions are made. But in neuronal science and psychological work on the unconscious there are as yet extremely few findings in this area that have achieved any sort of common consensus. Hence the only way these scientists know how people make decisions is by constructing choice theories out of their own introspection.

This attitude is untenable. Except where we have objective grounds for deeming participants' introspection to be inferior to our own, we must supplement our own introspection with that of others. In the words of Simon (1991a), we need to be wary of arm-chair theorising, and get empirical information such as self reports. This is particularly important for any of us already wedded to a theory, in that it may bias our introspection on how we ourselves decide to interpret our participants' motivations.

Further, even though neuronal science is in its infancy, progress has been sufficient to identify two things consonant with the conclusions we draw from self-reports. One is the role of secondary satisfactions in decisions, *eg* Damasio (1984) discerned the role of a secondary satisfaction, anticipations of fear, in making good decisions and Camille et al. (2004) discerned the role of another secondary satisfaction, anticipations of regret, in making good decisions. The other is that the complex calculations of multiplication and addition that we typically conceive as being used to compute the value of acts under EUT and other rank dependent procedures would need to be done in the conscious, so-called calculating, sections of the brain. If participants fail to report doing such multiplication and additions, this is *prima facie* evidence that indeed these operations did not enter their choices.

Some have proposed eg Duncan Luce (private email) that choosers obey EUT (or a standard rank dependent generalisation thereof) without doing these steps in equation (3.5). Eg wihtout doing any multiplication of utilities by their probabilities and then adding them up to get the value of an act. Instead it is proposed that they do it simply by obeying all the axioms in a particular axiomatisation. But this is to forget that the axioms are not operational in this sense. Most axiomatisations start for instance by assuming that the individual already knows his preference ranking of acts. This is Postulate 1 in the axiomatisation in Savage (1954), and a pre-axiom postulate in Neumann & Morgenstern (1947, 1972). Taking into account this axiomatic beginning, EUT consists in checking via all the other axioms (which put restrictions on this ranking). This is because the axioms involve hypothetical reasoning of the sort identified by Gilboa & Schmeidler (2001).

"... to apply EUT ... one needs to engage in hypothetical reasoning, namely to consider all possible states and the outcome that would result from each act in each state" [(Gilboa & Schmeidler, 2001, p43)].

The matter of checking for conformity with EUT (while dodging any operations of multiplication and addition) is, in an operational sense, endless. Black (1986) skilfully illustrates this for a case where it was only a ranking of books in a bookstore. At any time in doing the check of being consistent with EUT the chooser can be astonished that he is inconsistent with EUT. This was after all the astonishing dilemma of Savage, when in his famous Paris encounter with Allais, he discovered that his ranking disobeyed EUT. Savage did not know which pair of acts he had "wrongly" ranked, and after all he only had four pairs of ultra simple acts to rank.[1] But the axioms are strong. They have bite. It is simple to believe like Savage that one is a good EUTer and then discover an incompatibility on one's preference ranking with an axiom.

Of course the reverse does not hold, axioms can readily be demonstrated, using skilfully tiny choice sets, to be disobeyed by participants. The purpose of this paper however, is not to ascertain procedures *not* followed by participants. It is rather the positive one of ascertaining *how* participants do choose.

But what does all this amount to in terms of choosers' self-reports? To argue that any standard rank dependent choice theory such as EUT is obeyed because the chooser has obeyed all the axioms in a particular axiomatisation involves far too many steps to be done in the conscious part of the brain. To offer an exclusively axiomatic defence of any standard rank dependent theory is to declare it empirically uncheckable given the stupendous number of consistency checks required, in any experimental or other empirical setting. It also makes it obscure whether this class of theories can contribute to economic understanding and policy analysis. How do economists, untrained in the unconscious, discover the cardinal utility mapping unique apart from scale and origin, *postulated as formed in an identical way under all standard rank dependent theories*? (The the-

[1] To work this preference ranking out, he constructed his, as he put it "extralogical and loose" sure-thing principal. As Savage noted he left his sure-thing principle unaxiomatised because unlike EUT it was based on "knowledge" and "possibility", Savage (1954). However as shown in Pope (1991a), Savage's own sure-thing principle to clarify his preferences has the problem of irrationally truncating probability distributions. To rationally clarify preferences requires, among other things, a stages-of-knowledge framework.

ories differ only in their decumulative probability weights in aggregating these utilities.) It seems to leave it entirely a matter of faith that utilities extracted from empirical data and assumed to be those of EUT reflect, and it perhaps relates to the econometric observation that the utility mapping is the worst estimated entity in economic theory). Indeed there is an extreme amount of faith involved in believing that unconscious utilities guide choice in a manner according with EUT. Tversky's conjoint analysis (Tversky (1967)) reveals that no set of unconscious probabilities and unconscious utilities employed in a manner compatible with EUT could explain the choices of his participants.

This book has not denied that the unconscious enters our choices. Its agenda is however that of Simon (1989) in his case for looking into the reasoning process. While in the 1960s Tversky sought to explain the deviations of his participants from EUT by means of his own introspection (on loss aversion) our agenda is to look at conscious choices, those involving reasoning. Our agenda is to see how we might approach building conscious choice models, and to see what theoretical entities should form their basic building blocks. We see this as a promising approach for developing theories with reasonable explanatory and predictive power for use in understanding market and social developments and advising on government policies. To this end, we ignore the speculations associated with exclusively unconscious brain level accounts of how some choosers (those not discovered as disobeying its axioms) follow some decision procedure.

The final methodological objection to be addressed here, is our supplementing self-reports on why actual choices were made with self-reports on hypothetical choices. One line of defence is that three Nobel laureates, Allais, Kahneman and Simon, based the empirical material underlying some of their choice theories on hypothetical choices. Allais (1988) constructed his rank dependent theory from the answers to hypothetical questions proffered by de Finetti, Malinvaud and other eminent scientists. Kahneman (with Tversky) derived prospect theory (1979) from participants' hypothetical choices. As Simon observed (conversation with one of the authors), there are many situations in which one cannot (even probabilistically) impose the dire consequences or reward the superlative consequences on participants. They are dire because unethical, and superlative since beyond the experimenters budget. Yet it is important to get some handle on choices in these situations.

In summary, we treat self-reports on actual and hypothetical choices as fallible. But we deem it unwise for experimenters to rely *solely on their own introspection* influenced as it is by the theories and culture of their own scientific circle to pronounce on why choices are made, and what should enter normative, prescriptive and descriptive choice models, and to pronounce on the range of different contexts to which their experimental findings apply. We treat participants' self-reports on their own actual and hypothetical choices as a valuable supplement to our own fallible introspection.

Part II

Experiments

Chapter 11

Participants and Their Choices

11.1 Particpants

Participants comprised predominantly graduate students and faculty. They came from the University of Graz's Faculty of Economics and Social Sciences (located in Graz, Austria), and the University of Virginia's Darden Graduate School of Business Administration (located in Charlottesville, USA), and from its departments of Economics and Mathematics on the Charlottesville campus. Our first session was in May 2003 with 15 participants and some aspects of its findings are reported in Pope et al. (2003). The four sequel sessions used very similar set-ups. For dates, locations and the sex and academic status splits, see Tables 11.1 to 11.3.

Table 11.1: Experimental Sessions

Session	Location	Date	Number of Participants
1	University of Graz, Graz, Austria	May 2003	15
2	University of Graz Graz, Austria	Nov. 2003	18
3	University of Virginia, Charlottesville, USA	March 2004	42
4	University of Graz, Graz, Austria	April 2004	20
5	University of Graz, Graz, Austria	April 2004	18
All			113

Table 11.2: Number of Participants by Sex and Academic Status

	Female	Male	Total
Faculty	2	11	13
Doctoral Students	8	10	18
MBA students	4	36	40
Undergraduate students	18	23	41
Computer Manager*	0	1*	1*
Total	32	81	113

* amalgamated into the category faculty plus in subsequent tables

Table 11.3: Percentage by Sex and Academic Status Structure in Each of the Five Sessions

Session	female	male	total	faculty plus*	students doctoral	MBA	Under-graduate	total
1	27	73	100	60	0	40	0	100
2	44	56	100	0	100	0	0	100
3	7	93	100	12	0	88	0	100
4	55	45	100	0	0	0	100	100
5	33	67	100	0	0	0	100	100
all	28	72	100	12	16	35	36	100**

* includes in session 1 also one computer manager participant
** rounding yields a sum of the items below the total

11.2 The Scenario: Potential Monetary Payoffs

Participants were handed a sheet on the context (for its wording in the English language version see Appendix C). This asked them to state whether they would take out protection insurance against an attack which would annihilate a sum available to them later. Later was about 24 hours later in the case of the first two Graz sessions (one at the end of a two-day workshop, the other at the end of a two-day short course). In the case of the Charlottesville Session, about two hours later (end of the lecture which followed the experiment).

The sum subject to attack was fixed in each session, at €80 (€:EUR) in the case of the four Graz experiments, and $100 ($:USD) in the case of the Charlottesville experiment. At the USD/EUR exchange rate of March 2004, these amounts are roughly comparable. We know that exchange rates change without comparable changes in purchasing power. In terms of the wealth, disposable income and going wage rate of faculty and students in Austria and the US, and in terms of the actual purchasing power of the two currencies in respectively Graz and Charlottesville, the comparability will therefore be rougher. We can merely state that the amount at risk is more than a trifle, but well short of that which would endanger the financial viability of participants.

The cost of protection insurance was fixed in each session, at a cost of €30 in the first two Graz sessions, at $15 in the Charlottesville session, and at €12 in the last two Graz sessions. In all sessions participants were asked to indicate whether they would take out insurance at this cost. They were asked to make their choice six times, for six different alternative levels of risk, namely 0.1%, 0.2%, 1%, 5%, 10% and 20%. The risks were first described in terms of the number of days in which there had been an attack in the last 1,000. Then the corresponding percentages were given.

In all sessions, prior to making any decisions, participants knew the following. Each would place his or her name on a piece of paper in a box. After the answers to all questions had been collected, the papers containing the names would be shuffled in a box in front of the class by one of the experimenters. In front of the group, one participant would then select one paper from the shuffled box and read out the name of the selected participant. The experimenters would then confirm from the collected written answers for what risk levels the selected participant had chosen to pay insurance money. If the one selected had chosen to pay insurance for all six risk levels, that participant already knew at the end of the session the monetary sum to be received, ie, abstracting from doubts on whether they would survive till tomorrow and the insurance contract would be honoured, they had chosen to bring forward by 24 hours the time when their risk concerning their asset was resolved. If instead the selected participant had merely insured for some or none of the risk levels he or she would write the six levels of risk on pieces of paper and place them in a box. These six pieces of paper would then be shuffled by one of the experimenters, after which the selected participant would draw one paper out, give it to one of the experimenters who would then read out its risk level in front of all participants. Later, at the end of the whole experimental session, in the presence of all participants, the selected one would discover from a random device whether an attack had occurred. In the Graz sessions this random device took the form of selecting out of a set of shuffled matches in a box either a marked one (denoting an attack), or an unmarked one, with the number of marked matches corresponding to the risk. In the Charlottesville session the random device was that the one selected would open his or her laptop. He or she would then call up the random number generator programme that faculty member Matthias Hild used in his MBA lectures, enter the level of risk, and allow this programme to determine for the level of risk whether an attack occurred.

Our experimental method of analyzing responses to the risk of an attack and loss of an asset is subject to three (standard) laboratory features. First, participants may deem the probabilities smaller than those stated because they both have a smaller chance of any particular risk level being realized and a smaller probability of being the participant selected to receive the monetary consequences of the choices made. Amongst our participants, only 2% reported being subject to this effect, results consonant with the decisions in control experiments not subject to these probability diminutions reported in Starmer & Sugden (1991). A second feature is that our design does not put participants who choose not to commit to pay for insurance in the situation of walking out of the experiment with less assets than on entering it, merely at risk of missing out on an amount that they could have gained in the experiment. Amongst our participants only 4% of participants reported that their decision would have been different if subject to being at risk of losing pre-existing assets as distinct from at risk of missing out on an increment in assets. From Loewenstein & Adler (1995), they may have underestimated this. A third feature is that participants may not be completely confident that the experimenters will honour insurance contracts, nor completely confident of surviving to reap the benefit. However our experimental method for discerning the motivators of choice and how distinct motivators are employed in choosing among acts is substantially independent of the *absolute* scale of the probabilities and the extent to which behaviour differs with respect to the risk of missing out on potential gains versus incurring losses.

11.3 Choices Made

Choices made and reasons given are itemised for each participant for each risk level by session in Appendix A. In the text we confine ourselves to summary statistics. Our choice set-up (which has the same cost of protection for all six alternative risk levels), resulted in nearly all participants exhibiting a threshold risk level. Such threshold participants bought protection at all risks higher than this threshold, at no risk level below it. For 30% this threshold was never reached: they never insured. For 10% this threshold was a lower risk than 0.1%: they insured for every risk level. The remaining 6% (seven participants, one in session 1, one in session 3, one in session 4, and four in session 5) displayed non-monotonic behaviour of insuring at a lower risk level, but not at higher risk levels. Six of these seven were undergraduates. Such decisions may be due to capriciousness as reported in Abbink et al. (1999), namely participants seeking to deviate sometimes from their general behaviour in the face of risk, so that if they have been daring for a succession of possible eventualities, for the next in the succession, they are cautious (and *vice-versa*). While we did not include a question inquiring into capriciousness, none spontaneously identified capriciousness as a reasons for choice. Hence this non-monotonic behaviour may instead be due to misinterpreting how to enter choices, or misinterpreting the choices available. We excluded from the analysis therefore this 6% of our original sample, reducing it from 113 to 106.

Table 11.4 summarises the monotonic insurance decisions of the other 94% of participants. No systematic effects of sex or wealth as proxied by academic status are discernible by eye or using Fisher's exact tests. It can be seen that the cost of insurance (roughly double in the first two sessions compared to the latter three) had no obvious systematic impact on insurance decisions. The inter-session differences in propensities to pay appear to be related more to context.

11.4 Experiences of Terrorism and Preoccupations With Violence

Participants in a Norwegian internet experiment had a much greater willingness to buy protection for a given risk level in the case of terrorism than other contexts like car accidents, Hausken (2004). Like results are reported in US field questionnaires, Sunstein (2003), with the additonal information that desire for costly protection was rather independent of how small the risks were. These findings suggest that a terrorism context might elicit a higher propensity to pay for insurance. In the Graz sessions, no connection was made between the word "attack" and terrorism. But this connection was present in the pre-experiment advertising to enlist enrolees for the Charlottesville session - which moreover turned out to occur occurred a few days after the Madrid March 2004 terrorist attack.

However the connotation of terrorism in the Charlottesville session is not the only difference compared to the Graz sessions. Another major difference is national experiences and preoccupations with violence. A high proportion of Charlottesville participants came from countries that frequently experience terrorist attacks such as Israel, India and South America, and a few from the US whose citizens are rather preoccupied about protecting themselves from violence in general and terrorism in particular. By contrast, the Graz sessions comprised mainly Austrians who seem much less concerned about attacks of either a terrorist or a purely pecuniary motivation (and who have over the last 30 years experienced far fewer attacks on property than in those other countries).

The Charlottesville session's experiences with and preoccupations with violence can elicit two sorts of responses. One is for it to reduce the propensity to insure, especially at "unfair" odds. This can happen through individuals becoming hardened in the face of danger and thus suffering less secondary dissatisfactions like worry, anxiety, fear. It can also happen through individuals acquiring a determination "*not* to give in to terrorists" by paying protection money. This is how one participant in the Charlottesville session said he reacted, writing that he experienced regular terrorist attacks in his homeland and that not giving in to terrorists underlay their decision in our set-up to stay uninsured. But our findings suggest that he constitutes a minority of session 3 participants.

Our results, summarized in Table 11.4, suggest that for most in the Charlottesville session 3, its context of terrorism combined with national experiences and preoccupations with violence, dramatically increased the propensity to insure relative to Graz sessions 1, 2, 4 and 5 that were devoid of this set of national experiences and preoccupations. Thus the much lower percentage uninsured at all levels for session 3 held in Charlottesville relative to sessions 1, 2, 4 and 5 held in Graz is statistically significant lower percentage by many orders of magnitude below 1%. This is true both when session 3 in Charlottesville is compared with sessions 4+5 in Graz for which the cost of insurance was roughly comparable, and when the Charlottesville session is compared with all the Graz sessions,

1+2+4+5. At this level of analysis, therefore, context combined with national experiences and preoccupations is utterly decisive, overwhelming the determinants of choice in conventional theories such as EUT and standard rank dependent theories. We shall however later introduce other evidence on secondary satisfactions to suggest that context combined with national experiences and preoccupations does not explain the whole of this strikingly higher propensity of the Charlottesville session to pay for protection.

11.5 The Ordering of the Risks

The order of the six risk levels was ascending in the case of session 2, descending in sessions 1 and 3. There was a lower insurance charge and inter-country differences in session 3, so that it is difficult to use it for discerning order effects. For this reason we initially concentrated on differences in the insurance propensity between sessions 1 and 2 that were both held in Graz.

The ascending order of session 2 is correlated with a higher number of participants choosing to insure for five of the six risk levels compared to session 1, in which percentagewise and absolutely the cost of protection was identical. These two sessions differed in whether the risk levels were rising or falling. This suggests that the order of alternatives may play a temporal role, with participants envisaging a progressively safer or riskier future depending on the order. They may then form their overall impression on the need for improving their security by paying for protection at *each* risk level on the basis of the *last* risk level, that furthest to the right. This may relate to findings of Fredrickson & Kahneman (1993), Kahneman et al. (1993) and Kahneman (2000) that people pay more heed to final levels in a sequence of connected actual past events. There may be a like influence when, as in making choices, hypothetical future happenings are involved.

But the session 2 participants were exclusively doctoral students, whereas session 1 contained only faculty, MBA students, undergraduates and a computer manager. This higher propensity to pay for protection in session 2 relative to session 1 may therefore have stemmed from the different character of the participants. However the results of this exploratory

Table 11.4: Insurance Decisions of those with Monotonic Behaviour and the Impact of Context†

Session	Highest Percentage Risk for which Insured						Percentage Uninsured at all Risk Levels						
	20	10	5	1	0.2	0.1	female	male	faculty plus‡	doctoral students	MBA students	Under-graduate	Total
1	1	0	2	1	0	1	21	43	36	na	0	na	64
2	3	3	4	3	0	3	0	22	na	22	na	na	22
3‡‡	21	4	8	2	2	2	0	5	0	na	5	na	6*,**
4‡‡	2	3	0	2	0	2	26	26	na	na	na	53	53
5‡‡	4	0	0	2	2	2	7	43	na	na	na	50	50
4+5‡‡	4	5	1	0	0	6	43	58	na	na	na	52	52*
1+2+4+5	9	8	7	2	0	9	35	54	63	22	67	52	46**

† 94% of all participants. See text for details.

‡ includes one computer manager.

‡‡ Sessions 3, 4 and 5 had just under half the cost of protection (absolutely and relative to the amount at stake) of the earlier sessions, one that made paying for insurance with a 20% risk a "fair insurance", whereas even with a 20% risk, the cost of insurance, was "unfair insurance" in sessions 1 and 2. However, it is only in the case of the Charlottesville session, where the context of terrorism was mentioned in the prior advertising, that the absolute (and percentage) cost of insurance appears to have mattered, suggesting that it is a context effect, not an effect of the cost of insurance alone. There is no statistical difference between sessions 1+2 with more costly insurance, and sessions 3+4+5 with cheaper insurance.

* a statistically significant difference, 2-tailed, Fisher's exact test, orders of magnitude in excess of 1% between session 3 on the one hand and 4+5 on the other.

** a statistically significant difference, 2-tailed, Fisher's exact test, orders of magnitude in excess of 1% between session 3 on the one hand and 1+2+4+5 (all the other sessions), ie both those with double the insurance cost, and those with about the same insurance cost.

foray into the ordering of the risks indicates its possible significance in the terrorism context on which we plan a separate set of experiments.

To check on this we conducted two additional experimental sessions with comparable participants. They are comparable in the sense that all were at a like academic level, third year undergraduates enrolled in a game theory course, of a rather homogeneous Graz region cultural background also, and experimental participants on the same day (but in two different rooms for the two different experimental set-ups). Session 4 was given risks with an ascending risk order, session 5 a descending risk order. There was no discernible difference in their propensity to insure at any risk level. At least therefore in this context, by itself the order in which risk levels are presented does not appear to influence the decision to seek protection.

Chapter 12

Willingness to Pay Compared with Choices in a Dictated Choice Set

There were relatively low proportions choosing insurance even at our highest risk level of 20%. Further information on participants' risk taking propensities and whether participants conform with EUT, we obtained in sessions 2, 4 and 5. We did this by asking how participants might decide in choice sets offering a range of costs of insurance, or choice sets that they themselves achieved via negotiation. Unfortunately not all participants understood that they should answer this question set, so that the set of answers totals only 44, just under 80% of the 56 participants in these three sessions.

As discussed in Chapter 5 above, and summarised in Table 5.1 above, this relates to an earlier change in *knowledge ahead*, from not knowing the choice set to discovering eg via negotiation, at least some elements in it. After the initial set of answers had been handed in, we asked participants in sessions 3, 4 and 5 to nominate the price at which they would be willing to buy protection.

It is a logical error in reasoning on the part of us experimenters if we treat discrepancies between choices with a dictated actual choice set and those in which the participants must construct their own (counterfactual) choice set as necessarily due logical inconsistencies and ignore the role of the different state of *knowledge ahead* in the two situations. To avoid this logical fallacy, we here describe discrepancies in the decisions in two situations from what standard economic theory predicts a "disaccord" (as distinct from an "inconsistency"), decisions that conform in this regard as being in "accord" (as distinct from being "consistent"). See Table 12.1.

From Table 12.1 it can be seen that most participants were in accord with standard economic theory anticipations in the sense that these participants nominated prices below or at the cost at which they had rejected insurance as the price at which they would buy protection,[1] and nominated prices at or above the cost at which they had chosen insurance as the price up to which they would buy protection. Those in disaccord with standard economics anticipations may simply have made a conscious (or unconscious) re-evaluation of what protection is worth to them after answering the earlier questions.

In asking participants to nominate the amount that they are willing to pay for protection, we are asking them in effect to nominate an amount arbitrarily close to a certainty equivalent. This is the certainty equivalent of staying uninsured and facing risk. We did not in eliciting this amount offer participants the opportunity to indicate any range of indifference or uncertainty with respect to indifference. Yet when given this opportunity, participants typically state a wide range of indeterminacy in nominating certainty equivalents to risky alternatives, Albers et al. (2000). However, re-evaluations in a different mind-set and general indeterminacy in certainty equivalent may not be the only two factors underlying the 27% who deviated from our standard economics conception of the consistent answer.

---

[1]There is an ambiguity in how to classify those who choose not to purchase insurance at a given risk level, yet nominate the exact cost as what they are willing to pay. In our sessions, this concerned only two participants. The willingness to pay price is in disaccord with their having not insured - unless we note the boundary condition of being indifferent. The boundary can be included in both directions, insuring, not insuring, and note that 1) this indifference is what is assumed to exist under standard gamble methods of eliciting EUT utilities, and 2) in the case of continuity (assumed in EUT), is arbitrarily (infinitesimally) close to the certainty equivalents (willingness to pay insurance costs). Given 1) and 2) in conjunction with the principle of giving participants the benefits of the doubt as regards accord vis-à-vis ourselves, we classified these two as in accord in Table 12.1.

12.1 Choice Set Effects

More deviated in stating a lower willingness to pay for protection than the price they had actually accepted. Other experimenters have also found some discrepancies between the decision taken by participants when the price is dictated (ie the choice set is presented to them as a *fait accompli* and when asked to nominate their willingness to pay price, *eg* Coursey et al. (1987), Harless (1989) and Brown (forthcoming). As in our results, amongst participants exhibiting disaccord, there has been typically a preponderance of participants buying at a higher price when the price is dictated, than the willingness to pay price nominated. In our case, the percentage offering a higher price was a mere 2%, those offering a lower price a hefty 23%.

A propensity to buy a product depends in general on various factors. Frequently EUT trained economists inadvertently and mistakenly think that the number and type of act in the choice set is irrelevant. This inference is not a matter of logic. Rather, as Black (1986) and Sen (1993, 2002) demonstrate, it is an empirical and implausible EUT assumption. Similar comments hold as regards the EUT assumption that it is irrelevant whether the choice set is presented to participants as a *fait accompli* - as choice with respect to a non-negotiable cost of protection - is logically equivalent to a willingness to pay choice. They are logically equivalent under EUT, but it is an error in reasoning to mistake EUT assumptions for logical truths.

There are many factors, not mutually exclusive, that may contribute to a lower willingness to pay amount being nominated than that at which a participant took out insurance. A major factor is likely to be the different mindset in which participants are placed when having to construct their own cut off price at which the cost of protection is too high. This may induce the mindset involved in searching for the choice set by means of negotiations, and deciding when to stop negotiating, discussed in connection with Simon (1991b) in Chapter 5 above. Such a mindset may induce a negotiating stance in which the participant does not indicate the highest price at which he or she would buy (in our context insurance). Instead in a negotiation frame of mind, the participant may be indicating the price at which he or she would like to get the protection. Any such reframing of our willingness to pay as a wish price we may deem an ambiguity in how we put the question.

Table 12.1: The Percentage of Disaccord between Dictated and Chosen Choice Sets

Session: student type‡	*2: doctoral students ascending risks*			*4: undergraduate ascending risks*			*5: undergraduates descending risks*			*total : 2+4+5*		
Disaccord / Sex	*Female*	*Male*	*All*	*Female*	*Male*	*All*	*Female*	*Male*	*All*	*Female*	*Male*	*All*
Disaccord†,††	75	40	56	33	0	8	0	0	0	39	15	25
Disaccord: pay more	0	0	0	33	0	8	0	0	0	6	0	2
Disaccord: pay less	75	40	56	0	0	8	0	0	0	33	15	23
Disaccord+empties*,**	75	40	56	100	40	57	50	44	47	68	41	53

‡ Only session 2 (doctoral students), and sessions 4 and 5 (undergraduates) were asked their willingness to pay.

† Undergraduates are on Fisher's exact test significantly less likely to be in disaccord than doctorate students at t level of significance well in excess of 1%.

†† Males are at the 10% level 2-tailed Fisher's exact test less likely to be in disaccord than females.

* Once allowance is made for the higher number of partial and full empties in answers for undergraduates, reducing their opportunity to be in disaccord, there is no statistical significance in the slightly lower disaccord percentages of undergraduates (sessions 4 and 5) compared to doctoral students (session 2) under either a two-tail Fisher exact test or a 1-tail test for the theory that undergraduates tend to be more in accord.

*,** Once allowance is made for the higher number of partial and full empties in answers for undergraduates, reducing their opportunity to be in discord, there is no statistical significance in the slightly lower disaccord percentages of males compared to females under either a two-tail Fisher exact test or a 1-tail test for the theory that males tend to be more in accord.

Table 12.1 legend continued

Disaccord $= 100 \cdot D/(D+A)$
Disaccord pay more $= 100 \cdot D^{+}/(D+A)$
Disaccord pay less $= 100 \cdot D^{-}/(D+A)$
Disaccord + empties $= [(D+FE+PE)/(D+A^{'}+FE+PE)] \cdot 100$
where

D^{+} Participants in disaccord with standard economic theory anticipations by being willing to pay "too much". These participants stated a willingness to pay for insurance for at least one risk level above the amount for which they chose not to insure when the insurance cost (and thus choice set) was dictated.

D^{-} Participants in disaccord with standard economic theory anticipations by being willing to pay "too little". These participants stated a willingness to pay for insurance for at least one risk level below the amount for which they chose not to insure when the insurance cost (and thus choice set) was dictated.

D Total number of participants in disaccord for at least one risk level, $= D^{+} + D^{-}$.

$A^{'}$ The sub-set of A who stated a willingness to pay for every risk level.

FE Participants with a fully empty sheet on willingness to pay - who failed to provide willingness to pay answers for any risk level, inferring these questions non-applicable because they had insured at all risk levels.

PE Participants with a partially empty sheet on willingness to pay - who failed to provide some willingness to pay answers for those risk levels at which they had insured, inferring these questions were only non-applicable for risk levels at which they had chosen not to insure.

There is no evidence that whether the order of risks is ascending or descending influences the degree of discord. There is however a greater propensity for accord of males relative to females, and of undergraduates relative to doctoral students. This may indicate more thoughtfulness and readiness to re-think on the part of females and higher level students. The lesser disaccord might also relate to some difference in the oral presentation. Session 2 of doctoral students was conducted by a different co author from that of sessions 4 and 5 of undergraduates. The difference in oral presentation may have caused more of the undergraduates to misconstrue and think some or all of the willingness to pay questions were not applicable to them. But the difference in oral presentation may also be what caused most undergraduates who believed that the willingness to pay questions were applicable to them to infer that this question was not to be interpreted as what did as a "good" or "fair" or "desirable" price. On a two-tail Fisher exact test, the greater accord of the undergraduates is significant at a level decidely more stringent than 1%. The greater accord of males compared to females is significant only at the 10% level (regardless of whether a 1- or a 2-tail test is used, the right p-value being .08 and the 2-tail p-value .09).

These statistically significant differences in degrees of accord may however be pseudo. Quite a number of undergraduates partially or wholly failed to answer this counterfactual question set and thought it only applied to those who had chosen not to insure, or only to them in the cases in which they had not insured. We can control for the extreme case of this by augmenting the number in disaccord by the number who left answers wholly or partially empty, and thus imputing disaccord to all who left a blank for any risk level. When we do this there are no obvious differences as regards of whether the risks are presented in ascending order (sessions 2 and 4) or descending order (session 5), as regards whether participants are doctoral students or undergraduates, nor as regards whether participants are male of female.

12.2 EUT's Technique for Deducing Utility Shapes

A willingness to pay amount for protection is a statement of a certainty equivalent to that risk of retaining the asset. We can use the von Neumann & Morgenstern (1947, 1972) standard gamble technique. This allows us to infer from the boundary of participants' certainty equivalents (the price of insurance subtracted from the amount at stake), to the gamble of staying uninsured. This gamble has:

(i) a chance of retaining the full amount (1 minus the probability of an attack), and

(ii) a chance of losing it all (the probability of the attack).

If our participants use EUT to choose, we can therefore deduce the shape of their EUT utility mappings as follows. Participants asked this willingness to pay set of questions were from Graz. For them the full amount is €80, and losing it all is €0. The EUT scale is fixed apart from scale and origin. Set this scale to have a utility of zero for the minimum outcome, €0, and a utility of 1 for the maximum outcome, €80.

Then "utilities" deduced from EUT are on the vertical axis, and the outcomes on the horizontal axis as follows. From the EUT equation (3.5) of chapter 3 above, if there is a 0.2 chance of an attack, there is a 0.8 chance of getting a utility of 1 (from the €80), so it has a utility of 0.8, and the willingness to pay sum of €WTP for the 0.2 attack risk, yields a certainty equivalent of €80 - €WTP for a 0.2 attack risk. This gives an internal point on the EUT implied "utility" function that we may term point A. This is additional to our lower bound origin of (0, €0) and upper bound of (1, €80) set by construction under EUT and independent of the data.

We get additional internal points in the same way using the data from the participant. Thus from equation (3.5), if there is a 0.1 chance of an attack, there is a 0.9 chance of getting a utility of 1 (from the €80), so it has a utility of 0.9, and the willingness to pay sum of €WTP for the 0.1 attack risk. This yields an internal point we may denote B, namely the point at the coordinate on the x-axis of outcomes of a certainty equivalent of €80 - €WTP for the 0.1 attack risk which has on the y-axis of utility the value 0.9. To infer the utility function, we use linear extrapolation between points.

12.3 EUT's Modal Double Twist "Utility Function" Deduction

To illustrate computation of the internal points, let us take participant 5 in session 4. He has the most popular shape of a utility function. His EUT implied "utility" function is depicted in Figure 12.1.

Figure 12.1: The Modal Double Inflected EUT deduced "Utility" Function, a Primarily Convex Case in the Data Interval - Participant 5 in Session 4

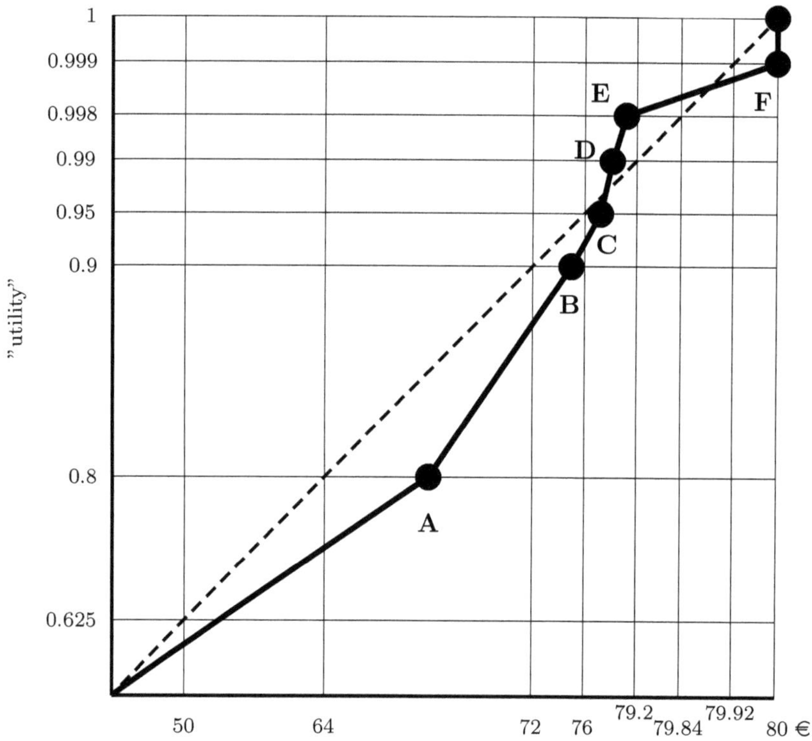

- - - - linear NM utility or fair bet price benchmark
——— participant 5 in session 4's EUT imputed "utilities" with linear interpolation between each imputed "utility".

Risk of attack	0.2	0.1	0.05	0.01	0.002	0.001
x denotes chose to insure						
WTP in €	5	4	3	2	0	0
Compared to linear NM utility or fair bet	<	<	<	>	<	<
Certainty equivalent: 80 - WTP	75	76	77	78	80	80
Classification	Convex then Concave then Convex-Vertical Consistency of WTP and insurance decisions					

It can be seen that it has two inflections, from at lower outcomes being convex, to then becoming concave, to then reverting to being convex - and so sharply convex that the curve is vertical. How does this shape appear out of the assumption that he obeys EUT? Consider first the point A which has a "utility" of 0.8 since from how we set the origin and scale, under the EUT assumptions, this risky act has a utility of 0.8 times 1 which is 0.8. To find the coordinate on the horizontal outcomes scale, we need to know the certainty equivalent. (Remember back to Chapter 3 where it is explained that EUT evaluates each outcomes "as if certain", Friedman & Savage (1948).

The certainty equivalent is the amount the person deems would give him enough utility to be willing to take that sum for sure instead of the risk of staying uninsured. Under the continuity assumption of EUT, this is arbitrarily close to the amount he would pay for the sure act of insurance instead of a 0.8 chance of €80. From the table beneath the graph in Figure 12.2, it can be seen this participant said he would pay €12 for insurance in these circumstances. This means that (€80-€12)=€68 is for him the certainty equivalent of a 0.8 chance of €80. This is the point A on Figure 12.2.

The point B on his EUT implied "utility" function is obtained by doing a like calculation for how much he is willing to pay for insurance if the risk is 0.1. This is €5. Then the coordinate on the utility axis is $1 - 0.1 = 0.9$, and the coordinate on the outcomes axis is (€80–€5)=€75.

The point C on his on his EUT implied "utility" function is obtained by doing a like calculation for how much he is willing to pay for insurance if the risk is only 0.05. This is €3. Then the coordinate on the utility axis is 1-0.05=0.95, and the coordinate on the outcomes axis is (€80–€4) = €76.

Like calculations yield the utility coordinates for D, E F, the other three positive risk of attack levels of respectively 0.01, 0.002 and 0.001. These correspond respectively to utilities of 0.99, 0.998 and 0.999 for which his willingness to pay was respectively €2, €1 and €0 so that their associated outcomes axis coordinates were respectively €78, €79, and €80. His utility function is the thick black line mapped out in Figure

12.2, while the linear EUT mapping, the "fair bet choice" for a benchmark is the dashed line.

Our data points range over low probabilities of loss from 0.2 to 0.001. This is because insurance contracts seldom involve probabilities of risk in excess of 0.2 and indeed often rare events, ones with probabilities much smaller than our smallest. The complements of these probabilities are the internal utilities on the graph. In reading the graphs note the non-linear scales chosen given that we lack internal points below a utility of $1-0.2=0.8$ - otherwise the plots would be unreadable - the space mainly occupied by the dataless interval of utilities between 0 to 0.8.

This most popular shape of convex then concave then so extremely convex as to be vertical, has multiple forms. In Figure 12.1, the participant's EUT implied "utility" function is primarily convex, convex until a utility of 0.998, and then tilts to be concave and then convex-vertical. In other cases of this shape, there is a greater concave range. Such is the case for participant 9 in session 2 depicted in Figure 12.2. This participant's EUT implied "utility" function is also convex then concave then convex-vertical. But it shifts from convex to concave at a lower utility, at a utility of 0.95 and remains concave until a utility of 0.999, and so is concave for about half the depicted utility points before tilting abruptly to convex-vertical.

12.4 EUT Implied Utility Shapes Classification and Summary

Table 12.2 summarises the shapes of our participants. It provides total statistics concerning concavity and verticality, and a total list of shapes in terms of concavity, convexity, linearity and verticality.

Figure 12.2: The Modal Double Inflected EUT deduced "Utility" Function, a case with concavity for half the data interval - participant 9 in session 2

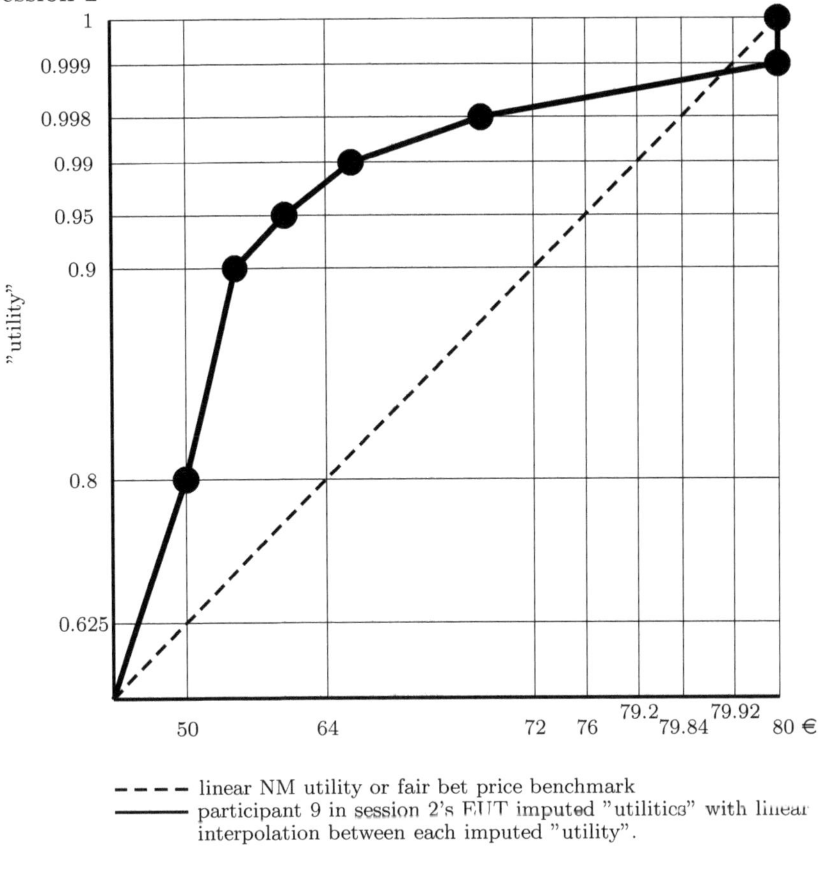

Risk of attack	0.2	0.1	0.05	0.01	0.002	0.001
x denotes chose to insure	x					
WTP in €	30	25	20	15	10	0
Compared to linear NM utility or fair bet	>	>	>	>	>	<
Certainty equivalent: 80 - WTP	50	55	60	65	70	80
Classification	Convex then Concave then Convex-Vertical Consistency of WTP and insurance decisions					

Table 12.2: EUT Implied Utility Shapes

Session‡	2			4			5			Total		
Shape / Sex‡‡	*F*	*M*	*all*	*F*	*M*	*all*	*F*	*M*	*all*	*F*	*M*	*all*
concave till after 5%	0	0	0	14	14	14	0	11	8	6	8	7
concave till after 10%	0	0	0	14	14	14	33	22	25	11	12	11
vertical segments*	88	80	83	71	100	86	100	67	75	83	81	82
vertical segment begins before 1%	38	0	17	29	29	29	0	11	8	28	12	18
Shapes Classification‡‡‡												
1 concave then linear	0	0	0	14	0	7	0	0	0	6	0	2
2 concave then convex-vertical**	13	20	17	14	29	21	66	11	25	22	19	20
3 concave then convex-vertical then concave	0	0	0	0	14	7	0	11	8	0	8	5
4 concave then convex-vertical then concave then linear	13	0	6	0	0	0	0	0	0	6	0	2
5 concave then convex-vertical then concave then convex-vertical	0	10	6	0	0	0	0	0	0	0	4	2
6 concave then convex then concave then convex-vertical	13	0	6	0	0	0	0	0	0	6	0	2
7 concave then linear then convex-vertical then concave	0	10	6	0	0	0	0	0	0	0	4	2
8 concave then linear then concave then convex-vertical	0	10	6	0	0	0	0	0	0	0	4	2
9 linear	0	20	11	0	0	0	0	0	0	0	8	5

‡ Only sessions 2 (doctoral students), and sessions 4 and 5 (undergraduates) were asked their willingness to pay and of these only 44 gave answers. See Figures 12.1 to 12.5 in this chapter and Appendix B for the implied EUT utilities of each participant.

‡‡ F denotes female, M denotes male.

‡‡‡ Each "then" describes a change in shape as the outcome increases. Convex-vertical means that the shape is not merely convex but that the right hand side of the convex shape is vertical.

* Incompatible with EUT.

** Mode of those beginning concave.

*** Mode of these beginning convex.

Table 12.2 continued

Session‡	2			4			5			Total		
Shape / Sex‡‡	*F*	*M*	*all*	*F*	*M*	*all*	*F*	*M*	*all*	*F*	*M*	*all*
Shapes Classification‡‡‡												
10 linear then concave then convex-vertical	13	0	6	0	0	0	0	0	0	6	0	2
11 convex-vertical	0	0	0	14	14	14	0	0	0	6	4	5
12 convex then concave	0	0	0	14	0	7	0	0	0	6	0	2
13 convex then concave then linear												
14 convex then concave then convex-vertical***	25	20	22	14	43	29	0	22	17	17	29	23
15 convex-vertical then concave then convex-vertical	13	0	6	29	0	14	0	0	0	17	0	7
16 convex then concave then convex-vertical then concave	0	0	0	0	0	0	33	11	25	6	8	7
17 convex then concave then convex then concave then linear	0	0	0	0	0	0	0	0	8	0	4	2
18 convex then concave then convex then concave then convex-vertical	0	10	6	0	0	0	0	0	0	0	4	2
19 convex then linear then concave then convex-vertical	0	0	0	0	14	7	0	0	0	0	4	2
20 convex then linear then convex then concave then linear	0	0	0	0	0	0	0	11	8	0	4	2
21 concave then convex then concave then linear	0	0	0	0	0	0	0	11	8	0	4	2

‡ Only sessions 2 (doctoral students), and sessions 4 and 5 (undergraduates) were asked their willingness to pay and of these only 44 gave answers. See Figures 12.1 to 12.5 in this chapter and appendix B for the implied EUT utilities of each participant.

‡‡ F denotes female, M denotes male.

‡‡‡ Each "then" describes a change in shape as the outcome increases. Convex-vertical means that the shape is not merely convex but that the right hand side of the convex shape is vertical.

* Incompatible with EUT.

** Mode of those beginning concave.

*** Mode of these beginning convex.

12.5 No Participant With Implied EUT Constant Relative Risk Aversion

If our participants not only obey EUT, but also conform with the popular assumption of constant relative risk aversion, so-called - an isoelastic concave mapping from "as if certain" money into utilities - these utility shapes should be everywhere concave. It can be seen that none is. Only 7% are approximately thus, approximately in the sense of remaining concave until the risk level falls below 5%. Insurance decisions are normally about small risks. Risks above 5% are very rarely insured. But even if we consider those that are concave till after risks of 10%, we still only have 11% of participants classified as concave. As a representative agent or typical potential insuree characterisation, then, Table 12.2 suggests that constant relative risk aversion is an unwarranted assumption even for risk levels larger than the norms for which insurance companies insure. Note moreover, that absence of concavity everywhere is here demonstrated not merely for high outcomes if present for small outcomes as in Rabin (2000), but demonstrated here within a tiny range of small outcomes, in the vicinity of €50 to €80.

12.6 Only 5% With Implied EUT Linear Utility

With respect to the other popular EUT assumption, especially in finance, of a linear "as if certain" mapping from money into utilities, the so-called risk neutral case, Table 12.2 shows us that only 5% conform.

12.7 The Typical Multiple Inflection Points in the EUT Implied "Utility" Functions

In Table 12.2 we see that the modal shape Figures 12.1 and 12.2, is shape #14, is the modal shape, namely convex then concave then convex-vertical. Figure 12.3 is of participant 17 in session 2, a representative of the second most popular shape, shape #2, namely concave then convex-vertical. Figure 12.4 is the tie third most popular shape, shape #15, namely convex-vertical then concave then convex-vertical. Figure 12.5 is the other tie third most popular shape, shape #16, convex then concave then convex-vertical then concave. This is illustrated for participant 1 in session 5. The implied utility functions for all the other participants in the willingness to pay questions are in Appendix B.

Figures 12.1 to 12.5 illustrate, and the summary of all the figure shapes in Table 12.2 reveal for nearly all participants a set of queerly scallopped shapes. In viewing the twisting scallopped shapes of the EUT implied utility mappings summarised in this table, bear in mind that our EUT elicited utilities comprise only six interior points in the utility scale, those corresponding to the six risk levels 20% to 0.1%. In between each of these there could be many more scallops in each participants' utility map. Appendix B graphs the EUT implied utility for each participant in the willingness to pay questions.

12.8 The Incompatibility of the Weird Implied Shapes with EUT Axioms

Some may wish to argue that twists in the utility shape is no evidence again EUT, merely against the simple shapes commonly assumed. Friedman & Savage (1948) after all postulated one twist. This however is implausible when we nest these findings within other experimental investigations eliciting the utility shape under the EUT standard gamble technique. When the full range in both the outcomes and the risks is used and the elicitations begin in the vicinity of the 50% risk of either the worst or the best outcomes, or a certain amount in the vicinity of the middle of the domain, we have benchmark findings. These are in the seminal meta studies of Hershey et al. (1982); Hershey & Schoemaker (1985) which showed that overlying minor twists in the standard gamble elicited utility mapping is an overarching twist from concave for low outcomes to convex for high outcome under the probability equivalent method, from a convex shape for low outcomes, to a concave shape for higher outcomes under the certainty equivalent method. This overarching twist is systematic and dramatic enough to discern even at the average session level of analysis, ignoring individual differences. Yet under EUT assumptions, the two implied utility shapes are identical since the two elicitation methods are logically equivalent! Pope (2004) shows how this overarching twist in the EUT utility mapping elicited arises from omitted variables bias, from the omitted role of negative secondary dissatisfactions from risk causing a bunching of probabilities away from the probability of death under the probability equivalence method. Conversely Pope (2004) shows that under this EUT certainty equivalent method the

Figure 12.3: The Popular Concave then Convex-Vertical EUT deduced "Utility" Function illustrated with participant 17 in session 2

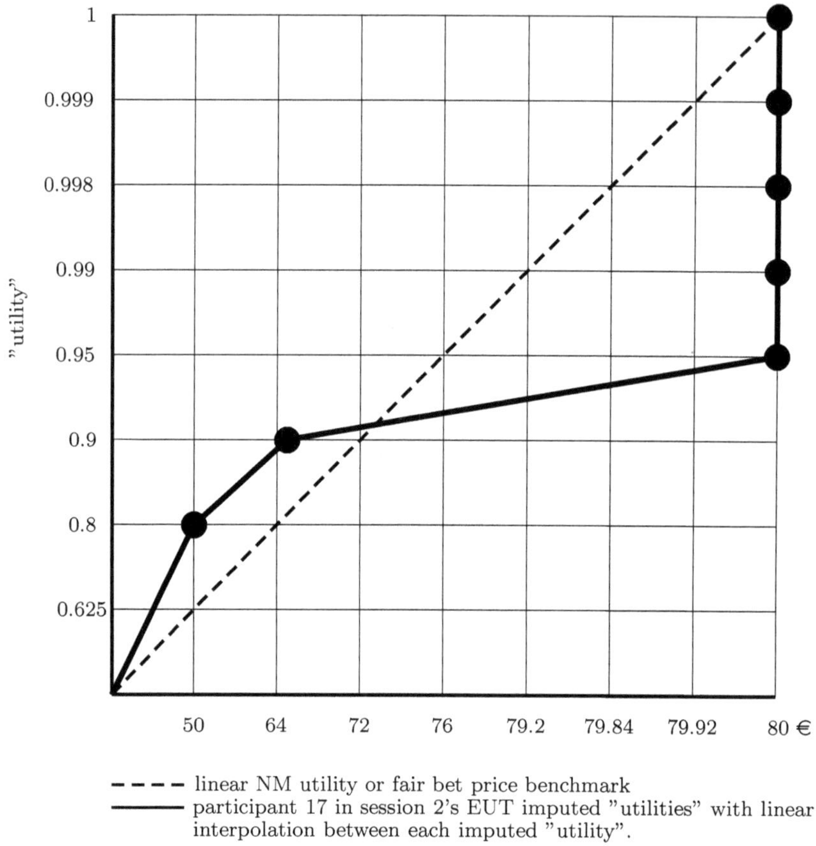

Risk of attack	**0.2**	**0.1**	**0.05**	**0.01**	**0.002**	**0.001**
x denotes chose to insure	x	x				
WTP in €	30	20	0	0	0	0
Compared to linear NM utility or fair bet	>	>	<	<	<	<
Certainty equivalent: 80 - WTP	50	65	80	80	80	80
Classification	Concave then Convex-Vertical Inconsistency of WTP and insurance decisions					

Figure 12.4: The EUT deduced "Utility" Function that is Double-Twisted with Two Vertical Segments illustrated with participant 8 in session 2

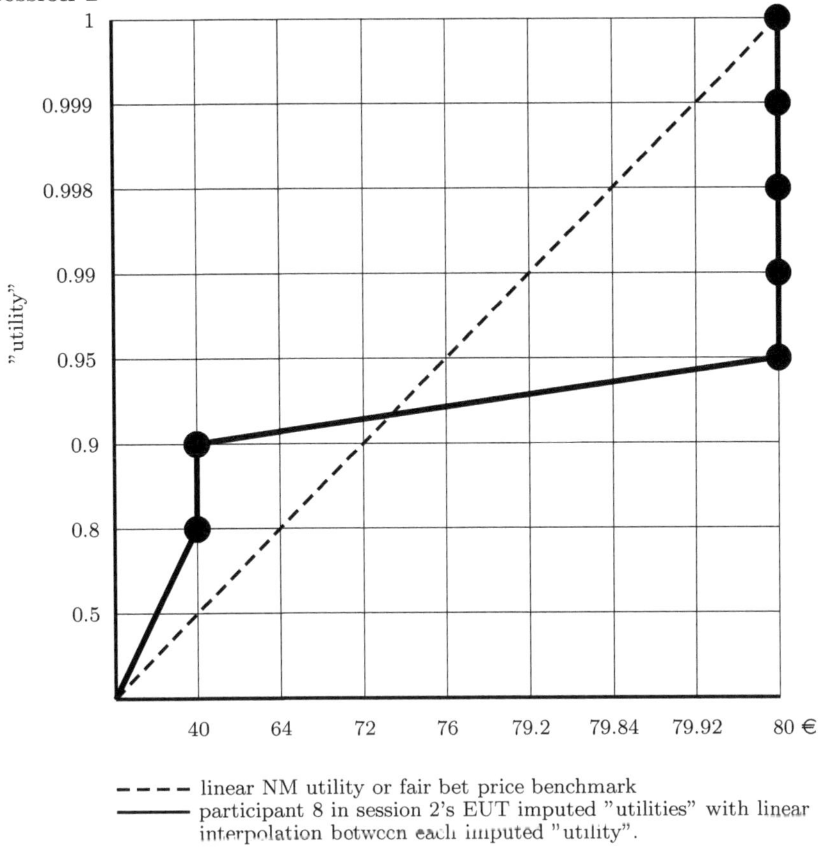

Risk of attack	0.2	0.1	0.05	0.01	0.002	0.001
x denotes chose to insure	x	x				
WTP in €	40	40	0	0	0	0
Compared to linear NM utility or fair bet	>	>	<	<	<	<
Certainty equivalent: 80 - WTP	40	40	80	80	80	80
Classification	Convex-Vertical then Concave then Convex-Vertical Consistency of WTP and insurance decisions					

Figure 12.5: The Triple Twisted EUT deduced "Utility" Function illustrated with the case of participant 1 in session 5

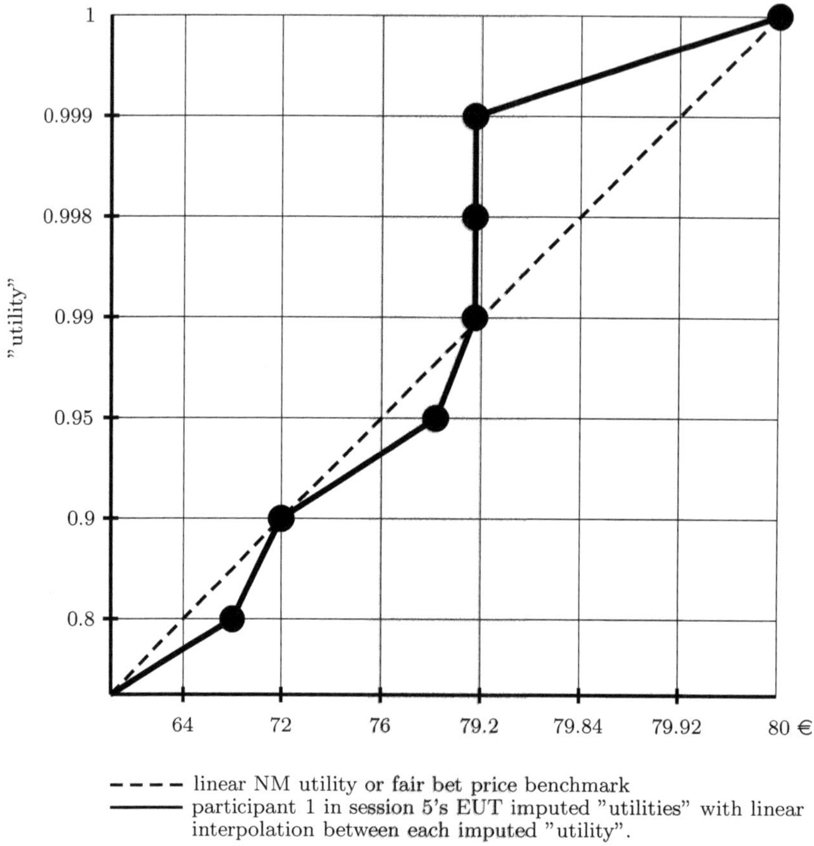

Risk of attack	0.2	0.1	0.05	0.01	0.002	0.001
x denotes chose to insure						
WTP in €	12	8	2	1	1	1
Compared to linear NM utility or fair bet	<	=	<	>	>	>
Certainty equivalent: 80 - WTP	68	72	78	79	79	79
Classification	Convex then Concave then Convex-Vertical then Concave Inconsistency of WTP and insurance decisions					

reverse overarching twist arises from omitted variables bias, from the omitted role of negative secondary dissatisfactions from risk causing a bunching in the certainty equivalents (willingness to pay) proffered for all sizable risks of death away from full health.

Our elicitations of utilities under the von Neumann and Morgenstern standard gamble technique for the three Graz sessions are within a narrow range of outcomes. The worst is to miss out on an asset increment of €80, not to lose one's life or entire possessions and so forth. Further the risks we use to elicit utilities in this range are also concentrated, from 80% to 99.9% chance of the good outcome (from 20% to 0.1% of the bad outcome). For the very confined range of interior outcomes and interior risk levels investigated in this paper, such predominant twists may not occur. The limited degree of risk and amount at risk renders the negative secondary dissatisfactions of a limited degree of knowledge ahead less dire. It can even be limited enough to cause the positive secondary satisfactions of thrills to dominate.

Furthermore, from Table 12.2, the vast majority of our participants, 68%, have more than two twists just in the small range of outcomes here being investigated. It is hard to imagine, and even harder to construe as plausible, that the real utilities of our participants wiggle around in the manner that the EUT assumptions underlying Table 12.2 imply. The multiple scallops, twists, wriggles are suggestive of omitted secondary satisfactions, both positive and negative, depending on the level of risk, suggestive that participants do not have a set of precise "as if certain" utilities for each outcome level, and that how they compare risky and sure acts in order to choose between them differs from the EUT procedure of adding up "as if certain" utilities by their probability weights.

For most of our participants, the multiple twists of their EUT implied utility mappings in this modest outcome range is not merely implausible, but irreconcilable with any parameterisation of these twisting scalloping EUT utility curve. This is because of the propensity of participants to include vertical segments. A full 82% do this, something that theory's axioms exclude. The four most popular shapes illustrated in Figures 12.2 to 12.5 have these vertical segments. This is partly a matter of threshold effects to which Allais (1952) drew attention when probabilities become smaller than five percent. Only 18% have vertical segments earlier than this. But the threshold effects cannot be incorporated in an EUT error theory such as Hey & Orme (1994) and Hey (2005), nor the Kahneman & Tversky (1979) theory of underweighting small probabilities. These vertical segments are a matter of participants treating sets of different risk levels as identical, a matter of thresholds deliberately chosen by participants perfectly capable of reading the decimal points distinctions. Some verticality is in the Albers et al. (2000) prominence theory and the Brandstätter et al. (2006) priorities theory, though not in the precise form here detected.

Chapter 13

The Role of Secondary Satisfactions

13.1 Method: Reported Reasons for Choice

By asking our participants why they chose, we hope thereby to obtain more robust results, more useful in prediction, than those concerning the protection cost individuals would choose in the very specific context and outcomes range of our protection experiment. The main technique employed has accordingly been to have participants explain why, for each of the six risk levels, they chose to either avoid or assume the risk of an attack. They had the opportunity to alter their choices while stating why they chose. This is because their choices and the reasons they stated were collected simultaneously (written respectively on the front and back of a single sheet). In cases where they chose to insure, they were asked to indicate if their reason was:

1 to avoid worry that in an attack they would lose it all and/or

2 \$85 is far better than \$0, but \$100 is only a little better than \$85 and/or

3 other reasons, with a please explain these.

In cases where they chose to bear the risk, they were asked to indicate if their reason was:

4 to enjoy the thrill that there might be an attack where I would lose all and/or

5 \$85 is only slightly better than \$0, but \$100 is far better than \$85 and/or[1]

6 other reasons, with a please explain these.

In the Charlottesville session the following questions were asked *after* the first set had been collected. Questions 7 and 9 entice them further to indicate use a rank dependent method of evaluation of acts. Questions 8 and 10-12 are on hypothetical situations. The last three questions were also asked in sessions 4 and 5, here too, *after* the first set of answers had been collected. In those circumstances in which you chose to pay for protection,

7 was a consideration behind your decision a greater weight than indicated by the simple probability weight for the utility from the *bad* outcome of an attack? If the answer is yes, explain how you weight the alternative possible good and bad outcomes.

8 would you have paid the \$15 for protection money if you had to pay it up front, in which case you would receive \$100 for sure later this afternoon?

In those circumstances in which you chose to not pay for protection,

9 was a consideration behind your decision a greater weight than indicated by the simple probability weight for the utility from the *good* outcome of no attack ? If the answer is yes, explain how you weight the alternative possible good and bad outcomes.

10 would you instead have paid for protection if tonight were the only night when you could take a special friend out to dinner that would mean a lot to you but which you could not afford to do unless you knew that tomorrow you would be at least \$85 richer than you are today?

11 would you instead have paid for protection if you wished to borrow money from a friend for a meal out tonight, and knew that this

[1]For the initial two sessions, the wording in the German-language version of question 3 allowed for a less dramatically convex, but nevertheless still exceedingly convex utility function, namely that "€50 is slightly better than €0, but €80 is also slightly better than €50".

friend would not lend you any more money today unless you could guarantee to repay it tomorrow?

12 would you instead have paid for protection if a friend or colleague would ridicule or blame you if they discovered later that you had not paid, that there was an attack and that you had missed out on $85?

13.2 Findings on Motivations

These are summarised in Table 13.1 below. Around 20% of non-undergraduates gave the secondary dissatisfaction of worry as a motivator for choosing the protection contract, and of these the majority gave avoiding this negative secondary satisfaction as their sole motivators for taking the contract, in all three population groups. For only 3% of undergraduates was this the case. The difference is weakly significant on Fisher's exact two-tailed test, at the 10% level, and for those who hold the prior hypothesis of undergraduates being carefree, significant at the 5% level.

The positive secondary satisfaction of enjoyable excitement was a motivator for choosing the protection contract in the case of nearly 30% of participants in the Graz sessions, with the majority of these giving this positive secondary satisfaction as their sole motivators for not taking out protection contracts. Perhaps because of reference to terrorism as the context of the attack in the pre-advertisement of the Charlottesville session, this attraction to chance in the terrorism context was not shared by that group where less than 5% nominated enjoyable excitement as a motivator for not paying for protection. This difference is significant under a 2-tailed Fisher's exact test well in excess of the 1% level.

Consider now the counterfactual questions 10, 11 and 12 asked in sessions 3, 4 and 5, on three other secondary satisfactions. Questions 10 and 11 introduced the scope for two instances of negative secondary satisfactions from being unable to commit and efficiently inter-temporally allocate their consumption flow. The vast majority answered *yes* when confronted with these consumption allocation inefficiencies: over 80% when it was being unable to commit to taking a special friend out to dinner, question 11, and over 70% when it was being unable to commit to repay for a meal for oneself, question 10. Question 12 introduced the scope for the negative secondary satisfaction of ridicule or blame in the event an attack and being uninsured. Over 20% stated that they would pay for protection if they anticipated such ridicule or blame ensuing.

In all five sessions, those indicating secondary satisfactions as a motivator for choice was nearly 80%. In all in the three sessions in which we asked participants about not merely two, but five instances of secondary satisfactions, over 95% gave secondary satisfactions from secondary satisfactions as a motivator for choosing whether or not to take out a protection contract. This in turn meant that the 99% confidence limit for the lower bound of the population proportion influenced by any of the above secondary satisfactions is 0.85, calculated using the exact binomial distribution. In turn, this lower bound will be an understatement of the percentage ever influenced by any secondary satisfaction in their choice. It is the lower bound for choice sets in which merely five distinct instances of secondary satisfactions were listed.

13.3 EUT Disobeyed

Nearly 80% of our participants are found to be influenced by secondary satisfactions and thereby disobey EUT, ie those who gave yes answers to questions 1, 3, or (in the case of the latter three sessions) to questions 10, 11 and 12. For the about one fifth of our participants not detected to be influenced by secondary satisfactions under our investigative methods, there are other requirements that need to be met before we can conclude that this fifth makes choices predictable under EUT. There needs to be a consistency with EUT of each such participant's answers under Questions 2 and / or 3 for any circumstances under which each such participant decided on a protection contract, and a lack of discrepancy with the implied utility function in their answers under Questions 5 and / or 6 for those circumstances in which such a participant decided against a protection contract.

Two participants for instance checked both Questions 2 and 5. This is reasonable if for each the satisfaction from each of the three asset levels is based on the probability of that asset level occurring, *ie* on the participant's different degree of *knowledge ahead* that he would encounter in the low and high risk circumstances associated with when he did and did not choose the contract. But checking both these questions is incompatible with EUT since it implies conflicting shapes for its *independent of knowledge ahead* utility function.

Again, EUT is not simply a matter of looking only at the shape of the utility function and ignoring the probability of occurrence of each outcome. Consistency with EUT is therefore likewise violated by those participants who checked Question 2 but did not check Question 3. To conform with EUT, they needed to check question 3 and state either their complete valuation formula, or at least mention that they formed their overall valuation of the risky act not to take a protection contract as inferior to the sure act of taking the insurance contract as follows. They added up their mutually exclusive utilities under this rejected risky alternative by multiplying the probability of no attack by the utility of receiving the full sum and comparing this with the utility of the guaranteed amount under the sure act, commitment to pay for protection. (If they did not impute to the zero outcome a zero utility, they also needed to mention multiplying the zero outcome by its probability and adding this on to get the overall value.) Yet no participant who checked Question 2 offered such an account in Question 3. Indeed nearly all who checked Question 2 left Question 3 blank, suggesting rather that they employed a single attribute procedure in choice.

Again, EUT is not simply a matter of paying for protection if probabilities of attack are higher, not paying for it if the probabilities of attack lie below some threshold. EUT involves the chooser comparing the probability weighted sum of utilities under higher and lower probabilities. Those therefore who checked Question 2 as the ground for insurance, should also have checked Question 3. In Question 3 they needed to explain either how the shape of their utility function was not so concave as to cause them to choose the contract at low probabilities, or else how they used all the individual axioms in an EUT axiomatisation to reach their conclusion. None of them did this however.

There were only 3 out of 101 participants who conformed with EUT in their answers. These were participants whose sole basis of choice was expected value, the special case in which EUT's *independent of knowledge ahead* utility function is linear. It is reasonable to assert that rather few of these three would conform in more complex realistic attack or terror contexts with larger sums at stake. This is because our laboratory set-up was specially designed to make EUT easy to obey.

Table 13.1: Secondary Satisfactions: Influences on Decision to Pay or Not Pay for Protection

Specific Positive or Negative Secondary Satisfaction	Percentage reporting this as a Motivator for Choice						
	females	males	faculty	doctoral students	MBA students	under-graduate	Total
Reasons for Paying for Protection							
To avoid a negative satisfaction							
Q1 worrying whether there will be an attack							
sessions 1+2	8	25	38	17	0‡	0‡	19
session 3	50	21	0	na	25‡	na‡	22
sessions 4+5	0	5	na	na	na	3	3
Q12 ridicule/blame for uninsured in an attack*							
session 3	67	17	25	na	21	na	21
sessions 4+5	21	32	na	na	na	27	27
sessions 3+4+5	29	22	25	na	21	27	24
To reap a positive satisfaction							
Q10 ability to commit for dinner with friend*							
session 3	67	91	100	na	88	na	89
sessions 4+5	64	79	na	na	na	73	73

* Only sessions 3 (faculty and MBAs), and 4 and 5 (undergraduates) were asked questions 10, 11 and 12.

** At the 99% lower confidence bound of the percentage influenced by any of the above secondary satisfactions (all sessions) is 67.1.

*** The 99% lower confidence bound of the percentage influenced by any of the above secondary satisfactions (sessions 3, 4 and 5) is 85. Note that this is higher than for all sessions because of the additional counterfactual questions asked on secondary satisfactions. These additional questions reveal other circumstances in which secondary satisfactions influence choice.

‡ The lower percentage of undergraduates influenced by worry is significant under Fisher's exact test, with a 2-tailed test at the 10% level, a 1-tailed test at the 5% level.

‡‡ The lower percentage influenced by thrills in Charlottesville, session 3, is significant under Fisher's exact 2-tailed test well in excess of the 1% level. On the role of its hint of a terrorism context, and of these participants coming mainly from countries experiencing or preoccupied with violence, see text.

Table 13.1 continued

Specific Positive or Negative Secondary Satisfaction	Percentage reporting this as a Motivator for Choice						
	females	males	faculty	doctoral students	MBA students	under-graduate	Total
Reasons for Paying for Protection							
To reap a positive satisfaction							
Q11 ability to commit to repay for own dinner*							
session 3	67	77	75	na	76	na	76
sessions 4+5	71	63	na	na	na	67	67
Reasons for Not Paying for Protection							
Q4 thrills/wondering about an attack							
sessions 1+2	25	30	13	28	33	67	28
session 3	0	6	0	na	6	na	6‡‡
sessions 4+5	29	32	na	na	na	30	30
sessions 1+2+3+4+5	25	19	8	28	9	33	21
Worry or Thrills (sessions 1 and 2)	33	50	50	39	33	67	
Any of the Above Secondary Satisfactions (all sessions)**	71	82	67	39	91	92	79
Any of the Above Secondary Satisfactions (sessions 3, 4 and 5)***	100	94	100	na	97	94	96

* Only sessions 3 (faculty and MBAs), and 4 and 5 (undergraduates) were asked questions 10, 11 and 12.

** At the 99% lower confidence bound of the percentage influenced by any of the above secondary satisfactions (all sessions) is 67.1.

*** The 99% lower confidence bound of the percentage influenced by any of the above secondary satisfactions (sessions 3, 4 and 5) is 85. Note that this is higher than for all sessions because of the additional counterfactual questions asked on secondary satisfactions. These additional questions reveal other circumstances in which secondary satisfactions influence choice.

‡ The lower percentage of undergraduates influenced by worry is significant under Fisher's exact test, with a 2-tailed test at the 10% level, a 1-tailed test at the 5% level.

‡‡ The lower percentage influenced by thrills in Charlottesville, session 3, is significant under Fisher's exact 2-tailed test well in excess of the 1% level. On the role of its hint of a terrorism context, and of these participants coming mainly from countries experiencing or preoccupied with violence, see text.

The set of monetary outcomes was known precisely, as was the set of associated probabilities and each of these involved only one significant digit. There was moreover for each act at most one outcome with a positive probability, and that outcome contained at most two digits.

This set-up does two things. One is to render it feasible for the even fractionally arithmetically competent to compute expected values: only one product in each case, and then only multiplying a single number by at most a two digit one. (However we would not suggest that the participating decision sciences faculty and graduate students were only fractionally arithmetically competent, or lacking a pocket calculators or laptops in the laboratory.)

The other and much more important thing that such a set-up does is to relieve participants of needing to ponder how much they might have at risk or the risk of attack. These we specified for them. In actuality the numerical risks are missing from most marketplace insurance contracts. Under EUT those choosing to accept a contract are assumed to have estimated these numerical risks themselves.

The difficulty for even eminent businessmen and scientists to construct a) their outcome space and b) a probability distribution over it may be judged in three ways. First, the observations of that keen observer of businessmen, Knight (1933) contended that business profits came from the difficulty of judging both a) and b), Langlois & Cosgel (1993). Second, experimental evidence supports Knight on the inability to formulate consistent probability distributions over outcomes, Slovic et al. (1982), and to furnish consistent utilities to individual outcomes, Borcherding & Winterfeldt (1988); Borcherding et al. (1991, 1995). Third, even the eminent attempt of Savage (1954) to offer axioms consistent with chooser's judging a) and b) while obeying EUT failed. His axiomatisation yields arbitrary results, so that it is only recently that success has been achieved in offering a feasible means of doing so, Karni (2006), though whether even Karni has succeeded is disputed by many. It is reasonable to say therefore that in the actual more complex realm under which protection contracts are offered, not even that three percent, will choose obeying EUT.

Chapter 14

Algorithms Versus Global Maximising Procedures

14.1 Choice Motivators

Our experimental evidence on reasons for choice accords with contracts being decided upon by looking at a set of distinct factors anticipated to yield positive or negative satisfactions. These factors and decision procedures reported by our participants included:

1 expected net value,

2 the increment in value of the amount possibly receivable over that guaranteed by paying for protection (*ie* those who checked questions 2 or 5),

3 specific secondary satisfactions, thrills, worry, planning inefficiencies, inability to commit, blame,

4 the probability of an attack,

5 the costs of protection,

6 the amount at risk,

7 comparisons of subsets of 4 to 6 (without mentioning whether these involved subtractions or ratios when in like units).

8 whether the insurance premium must be paid up-front.

There was considerable individual variation in the motivations and procedures reported. For details by sex and academic status and situations of paying, not paying for protection, see Table 14.1.

14.2 The Minority Using Expected Net Value

When all the numbers are simple and at hand, as in our scenario, some participants engaged in multiplication of primary satisfactions (measured by net monetary outcomes) with their probabilities. They then compared this magnitude with that of another operation, subtracting from the gainable sum the costs of protection. This expected net value calculation, was performed by 9%, ie 15 of 101 participants. Of the 15 using expected value, 12 were in session 3, the Charlottesville group comprising mainly MBAs. This is about double that typically found in experiments comprising undergraduates.

Charlottesville was one of the three experimental sessions in which the costs of protection was below the expected gross return on insuring. Hence the propensity of this set of participants to use expected value may account for the exceptionally high number of Charlottesville participants insuring for at least one risk level. Participants Sessions 4 and 5, like the Charlottesville session, had a low enough costs of protection to warrant insuring at the 20% risk level according to the expected value criterion. Yet not a single one in these two sessions of undergraduates used this criterion. In turn this casts some doubt on whether the Charlottesville session's distinctive context of terror and participants experiencing violence was the *only* factor leading to the markedly higher percent insuring compared to sessions 4 and 5. It suggests that an additional factor contributing to a high propensity to insure was the MBA usage of the finance expected value rule.

The 12 participants from Charlottesville in session 3 who reported using a net value calculation, all reported being swayed, in many cases decisively, by secondary satisfactions, ie reported disobeying EUT which excludes decision making based on a mix of expected value and secondary satisfactions considerations. This leaves the three participants of sessions 1 and 2 who used expected value and did *not* report being swayed also by thrills in deciding not to insure, as the only participants acting in accord with the standard finance assumption of EUT with linear utility enabling the expected value criterion. It was only in the later sessions 3, 4 and 5 that secondary satisfactions arising from planning and commitment difficulties and blame potential were introduced. When these were introduced, all those who calculated expected value criterion reported that it was not their sole criterion. This leaves open the likelihood that in more general scenarios these three participants from sessions 1 and 2 might also disobey the expected value criterion.

14.3 Non-usage of EUT With a Curved Utility Function

In most economic theorising and empirical work, an EUT chooser with an everywhere concave EUT utility function - indeed typically with a constant relative risk aversion EUT function - is assumed. But not a single participant reported using EUT with a non-linear utility function. Ie none used a concave or convex or more complex non-linear EUT function as their basis for contract choice. This is despite many participants being acquainted with EUT, including essentially all in the Charlottesville group, session 3. Perhaps many were in the same position as a participating mathematics professor in the Charlottesville session. He reported that he lacked the time to estimate the shape of his utility function and use it in reaching a decision on protection contracts.

This non-usage of non-linear EUT functions also accords with the evidence in the three sessions asked their willingness to pay for protection. Translated into implied EUT "utility" shapes, as explained in Chapter 12 above, only the linear shapes of 5% of participants could be said to be simple or plausible. None of the implied shapes were all concave or all convex, and over 80% outside any shape compatible with EUT. The investigation revealed in short, that the implications that people obey EUT and thus have an EUT utility function (ie one independent of knowledge ahead) leads to contradictions. It is thus unsurprising that when asked about why they chose, no participant reported usage of EUT with a curved utility function. Ie none reported multiplying his non-linear utility function derived value of the asset by the probability of their not being an attack and comparing it with his non-linear utility function derived value of the asset minus its protection costs.

Many reported considering the shape of their utility of money function (*ie* those who checked questions 2 or 5). But none reported putting a number on this value, and then doing a probability utility multiplication as required if they are to use EUT.

14.4 Minimal Role for Rank Dependent Generalisations of EUT

There is a notion that while EUT is disobeyed, all that is needed to obtain a good descriptive and predictive theory is the generalisation of its rule that mutually exclusive possible utilities be weighted by their probabilities of occurrence. Instead, as proposed in Allais (1952), the weights of utilities can be more general functions of their (de)cumulative probabilities than is EUT. Standard rank dependent theories retain three features of EUT.

First the atemporal (de)cumulative weights are assumed to be choice set independent. Second values of outcomes are assumed to stem exclusively from anticipated primary satisfactions, ignoring all anticipated secondary satisfactions. In this regard the entire class of standard rank dependent theories is disobeyed by some four fifths of our participants. These are the four fifths who reported being influenced even by the narrow range of secondary satisfactions here introduced.[1]

Third such theories share with EUT its procedure of forming an overall valuation of each act by forming (decumulative) probability weighted sums of the primary satisfactions obtained from the outcomes. None of the answers proffered by our participants to the first set of questions gave any indication of participants using even EUT's simple probability weights except when that theory is simplified down to linear primary satisfactions in money. Such persons refuse unfair insurance contracts. Those furnishing protection normally have costs and thus issue unfair insurance contracts. If the demand for insurance were limited to those with linear primary satisfactions in money, there could be no insurance companies and long run financial nonviability for governments providing terror protection or any other sort of insurance - unless potential insurees

[1] It might also be mentioned that insurers have to care about secondary satisfactions, the dispersion of their earnings, in order to issue genuine insurance contracts. Yet such caring conflicts with EUT and EUT+, Pope et al. (forthcoming).

had subjectively such higher risk probabilities than the insurers that the contracts unfair in the eyes of the insurers were fair in the eyes of the insurees.

Nevertheless we deemed it desirable to investigate further whether the demand for protection contracts at unfair rates might arise from protection purchasers computing (decumulative) probability weighted sums of satisfactions from alternative possible utilities, ie whether simple EUT or with a more complicated set of weights as in its rank dependent generalisations applied. Questions 7 and 9, added to the Charlottesville session, do this.

14.4.1 Minor Usage of Conservative Atemporal Aggregation Weights

Question 7 (see 13.1 above) inquired if decisions to pay for protection were influenced by giving a higher weight to the bad outcome of an attack than that indicated by the probability of an attack. Such conservatism of a greater atemporal aggregation weight on the bad outcome is hypothesized to be the case in most rank dependent theories, including the anticipated utility theory of Quiggin (1982, 1993), the invariance theory of Allais (1988), and cumulative prospect theory as in Tversky & Kahneman (1992) and Bleichrodt et al. (2001). A minority checked this question box. But of the explanations given by each participant for checking this box (see Tables A14 to A17 in Appendix A), only one was broadly in accord with rank dependent theories. For the choice sets, context and framing here investigated therefore, the evidence is of a tiny minority who conservatively atemporally weight. It may well be as in a footnote in Allais (1988), that such conservatism holds when the worst outcome is very dire. But evidently such conservatism does not hold in generality claimed by all these rank dependent theories, namely no matter what the worst outcome is.

Further the single participant unambiguously reporting such conservatism did not obey any standard rank dependent theory. Obeying would require also checking the conservative weighting box when reaching decisions not to take out insurance for other risk levels, and this was not the case. Further this participant reported being influenced by secondary satisfactions, something not permitted under these standard rank dependent theories.

14.4.2 Minor Usage of Aspiring Atemporal Aggregation Weights

Question 9 (see 13.1 above) inquired if decisions to take the risk and not pay for protection were related to the aspiration rank dependent theory of Lopes (1987), Lopes & Oden (1999). This theory allows some individuals to be more aspiring and thus to place an atemporal aggregation weight on the good outcome in excess of its probability. A minority ticked this question box. However of their requested explanations (see Tables A14 to A17 in Appendix A), at most two could be interpreted as stating that they put a higher atemporal aggregation weight on the good outcome. For the choice sets, context and framing here investigated therefore, the evidence is of a tiny minority who aspiringly atemporally weight. It may well be that were the addition to assets from risk taking much more massive than in our experimental scenario, an aspiration weight plays a role for a sizable proportion of the population. But such aspiring weighting is apparently not choice set independent. Further, both participants who unambiguously aspiringly weighted also reported being swayed by something excluded in the Lopes theory, namely being swayed by secondary satisfactions.

In short the above evidence from our participants points to implanting conservative or aspiring atemporal aggregation weights in decision models that differ from the standard rank dependent theories. Instead such weights need to be embedded in theories that (i) allow secondary satisfactions; (ii) are context and choice set sensitive; and (as discussed in the next section), are more heuristic, algorithmic.

14.5 Non-Numerical Treatment of Conflicting Motivations

For the majority of our participants what integration there is of conflicting motivations/goals is apparently less of a calculating procedure than the multiplication and addition of numbers found not only in EUT but also in most of its rank dependent generalisations. The Albers et al. (2000) prominence theory contributes handsomely on the role of choice sets in how fine are the numerical distinctions that we introduce into our decision processes. But it involves addition of numbers for which we find little evidence.

Nearly all our participants seem to be non numerical in what integration they do of the distinct sources of primary and secondary satisfactions arising out of the set of mutually exclusive possible outcomes of a risky act. For nearly all our participants, any integration done of distinct sources of satisfactions and of mutually exclusive possibilities is qualitative not qualitative. Participants do not mention whether comparisons take the form of subtractions or ratios.

When distinct satisfactions are not systematically integrated into overall indices of the valuation of whether or not to take out a contract, some other rule is needed to reach a decision and stop searching further for the best thing to do. That proposed by Simon (1955) in general terms has been made more concrete in the aspiration adaptation models of Sauermann & Selten (1962) and Selten (1998), and in the take-the-best models of Gigerenzer & Todd (forthcoming) and Brandstätter et al. (2006). Our experimental results suggest that these sequential non integrating approaches are promising.

14.6 Framing, Context and Choice Set Impacts on Heuristics

The probability of an attack is a key factor determining whether to pay for protection in our particular framings, contexts and choice sets. Our experimental results pertain to modest, though not insignificant sums at stake. This limits the proportion who sought a contract because of worry. For larger sums at stake, the experimental work of Schade et al. (2004) on the theft of valuables (not in a terrorist context) suggests that the proportion in the population suffering from secondary dissatisfactions such as worry would be higher than the 20% in our three sessions.

Probabilities entered the reasoning of many of our participants. As the probabilities of an attack were tiny, in the situations investigated, this limited demand for very "unfair" insurance. For risks including physical injury, in both terrorism and non terrorism contexts, the epidemiological and meta-experimental analyses of Sunstein (2003) indicate that where physical injury is the issue, probabilities play a more minor role than in our context where a financial asset is the issue.

The attack probabilities that deterred most of our participants from paying for protection ranged from low to very low. This is not supportive of Kahneman & Tversky (1979) and other studies used to infer that people inflate low probabilities. Our participants' reports hint at the theory of over-weighting of low probabilities being choice set dependent in some situations or non existent in others.

It might be thought instead that the decisions and reported decision procedures of our participants support the potentially different role of small probabilities in the priorities heuristic theory of Brandstätter et al. (2006). This however is not the case either, as follows. Our participants give a more prominent role to a), the probability of the bad outcome (an attack and losing all) than to b), the outcome that is worse by more than 10% relative to the next worst outcome (committing to insure and paying the costs). Thus well over 25% list a), the probability of gaining nothing as a grounds of choice, but none list b), the discrepancy between the worst outcome of nil under the act of staying uninsured and the worst (and only outcome) out of choosing to insure as a grounds of choice. Yet since the difference in worst outcomes is more than 10%, b) should decide the issue under the Brandstätter et al. (2006) priorities heuristics, and all should insure under all risk levels. Instead very few insured under any risk level, and not a single participant reported a procedure according with this prioirites heuristic.

Our participants' decisions and reports on why they chose support the general priorities heuristics agenda, and in this sense the pathbreaking work of the Austro-German Brandstätter-Gigerenzer team in demonstrating that such heuristics predict better than other decision rules in many choice sets, contexts and framings employed in economics and psychology experiments. Our findings demonstrate however that economists and psychologists, even those seeking new more realistic decision theories such as Brandstätter et al. (2006), have concentrated, like their predecessors such as Kahneman & Tversky (1979), on rather too narrow a set of choice sets, contexts and framing. This narrowness prevents scientists discerning the lack of universality of any (practical) decision theory and the readiness of decision makers to be swayed by factors drawn to their attention.

Eduard Brandstätter suggested (private email) that this conflict with the Brandstätter et al. (2006) priorities might due to the prominence of six different risk levels in our particular attack choice sets. Now every choice set involves framing and context with respect to the presence or absence of numbers for outcomes and likelihoods, the abstractness or concreteness, the gravity or frivolity, and the emotionality of the scenario, and the similarity or contrast in the alternatives available. It is infeasible to introduce choice without the scenario having all these dimensions. Each scenario as regards framing and context renders some aspects more prominent, not always those the experimenters anticipate!

Albers (2001) had demonstrated the choosers have rather less of the fixed check list and set of priorities proposed in the Selten-Gigerenzer-Brandstätter models in reaching choices. Our participants' decisions and reports corroborate such choice-set sensitivity. Indeed our findings are limited in their domain of application - limited to broadly similar choice sets, framings and contexts. Our book's finding in this respect accords with marketing techniques. These would be irrelevant if decision procedures are rigid and fixed because people are so clear on their goals and on their priorities within these goals!

Our book's findings disconfirm claims of universal validity for the algorithmic / heuristic procedures here analysed. Within SKAT our book's findings support the research agenda advocated by many scientists, including Kliemt (2004) and Hogarth (2006) that psychologists and economists investigate what we here term switching models. By switching models we mean models of the *changes* in people's key motivations and associated decision algorithms as choice set, framing and context *vary*. A small contribution in this direction is as follows.

First, our findings likely understate the key role of the secondary dissatisfaction that worry plays in some situations. Our experimental results pertain to modest, though not insignificant sums at stake. This limits the proportion that sought a contract because of worry. For larger sums at stake, the experimental work of Schade et al. (2004) on the theft of valuables (not in a terrorist context) suggests that the proportion in the population suffering from secondary dissatisfactions such as worry would be higher than the 20% in our five sessions.

Having a rather substantial sum at stake is not the only factor that can convert worry from influencing only 20% of participants as in our set of scenarios, to influencing high proportions to insure. Eg the *historical* context of having been debriefed after an earlier bad event can increase the amount of worry that people have. Thus dwelling on past traumatic events by debriefing motor vehicle accident victims was found in a controlled experiment on admissions to Radcliffe Hospital Oxford to increase worry (and as a consequence decrease health), Del Mar (2002). In turn this may increase their anticipations of being worried if they do not insure and enhance their propensity to insure and generally choose safer acts. Indeed we have hints of a related effect operating in session 3 where more insured and more reported doing so on account of worry, as discussed in chapter 11.

Second, our findings likely overstate the role that probabilities play in some situations. Probabilities entered the reasoning of many of our participants. As the probabilities of an attack were tiny in the situations investigated, this attention to how *tiny* the probabilities were, limited demand for even "fair" insurance.

However for situations involving physical injury, in both terrorism and non terrorism contexts, for many people, the Blatt threshold probability theory is false. The epidemiological and meta-experimental analyses of Sunstein (2003) indicate that for many people probabilities, and thus probability thresholds, play a more minor role once what is at risk changes from assets to people's health and lives. In some risk contexts, the population can become so alarmed that the smallness of the probabilities becomes in effect irrelevant. This irrelevance of their smallness is not because, as proposed in numerical overweighting of small probabilities theories, these probability numbers are inflated to other numbers by a particular functional form. It is rather, as Sunstein (2003) discovered, because in situations involving the risk of physical injury on a mass scale, many people pay no attention whatsoever to the probability magnitudes.

One of the reasons for people paying less attention to probabilities is the typical: (i) lack of agreement on what is the probability of a bad outcome; and (ii) lack of a shared understanding of what a probability means. In our scenario, we took great care to use devices that allowed all participants to (i) believe each probability number of the risk of an attack, and (ii) have a shared understanding of what that number means. In few real life choice sets is this the case. Care is required in inferring that probabilities of attacks play as big a role in other situations as they do in our particular insurance scenarios.

It must also be borne in mind the diversity of motivations: Table 14.1 reveals a diversity of algorithmic procedures used. This diversity is present even for our participants facing identical choice sets, framings and contexts. In this way, our participants' decisions and reported procedures disconfirm the existence of a universal decision procedure, *even for a single situation.*

Table 14.1: Reported Choice Motivator Percentages

Reasons for Paying for Protection	females	males	faculty	doctoral students	MBA students	under-graduate	Total
In accord with EUT							
Expected value alone							
Sessions 1+2	0	15	13	6	17	na	9
Sessions 3+4+5	0	0	0	na	0	0	0
Sessions 1+2+3+4+5	0	4	8	6	3	0	3
In Disaccord with EUT							
Utility shape but not in conjunction with EUT probability weighting of alternative utilities to value risky acts or in conjunction with secondary satisfactions							
Sessions 1+2+3+4+5	75	41	50	67	50	42	50
Greater atemporal aggregation weight than its probability on the worst outcome§							
Session 3	50	23	25	na	20	na	23
Greater atemporal aggregation weight than its probability on the best outcome§§							
Session 3	0	14	25	na	11	na	14
Secondary satisfactions‡							
Sessions 1+2	25	50	50	39	33	67	44
Sessions 3+4+5	100	94	100	na	97	94	96
Sessions 1+2+3+4+5	71	82	67	39	91	92	79
Expected value + secondary satisfactions							
Sessions 1+2	0	10	13	0	17	na	6
Sessions 3+4+5	0	19	50	na	25	0	14
Sessions 1+2+3+4+5	0	16	25	0	24	0	12
The probability of an attack only*							
Sessions 1+2+3+4+5	25	25	42	33	29	9	25

Reasons for Paying for Protection	females	males	faculty	doctoral students	MBA students	under-graduate	Total
In Disaccord with EUT							
Non-numerical comparisons with other motivators of:							
a) costs of protection**							
Sessions 1+2+3+4+5	11	19	25	0	26	12	17
b) amount at risk***							
Sessions 1+2+3+4+5	4	11	17	6	11	6	9
Difference in worst outcome with and without protection****							
Sessions 1+2+3+4+5	0	0	0	0	0	0	0
Independent of EUT							
Still insure if pay premium up-front							
Session 3	100	60	50	na	60	na	60

‡ Sessions 1+2 are separately reported because they did not receive questions 10, 11 and 12 on secondary satisfactions.

§ The hypothesis in most rank dependent generalisations of EUT.

§§ A hypothesis in the Lopes Aspiration Theory generalisation of EUT.

* In addition one participant listed the probability of an attack in conjunction with other factors.

** Five participants listed the costs as the sole motivator.

*** In addition one participant listed the amount at risk as a the sole motivator.

**** The Brandstätter et al. (2006) hypothesis. It is evidently valid only for a different class of choice sets.

Not listed in the Table:

- Three participants reported that they wanted to take the risk of getting "all or nothing".
- Two participants said that they are "risk loving". With respect to the partition of motivators into primary and secondary, this is ambiguous. "Risk loving" can mean eg (i) convex EUT utility function in conjunction with simple probability weights for determining the value of an act, (ii) rank dependent weighting of the mutually exclusive outcomes with more weight on the good outcomes , or (iii) secondary satisfactions - liking not knowing ahead. See Pope (1995) for the means of consistently distinguishing between (i), (ii) and (iii).
- One participant gave a pseudo reason- simply stated chose act with higher value, not *why* being uninsured has a higher value.
- One participant was optimistic about the attack and stated "I will have luck" as his only reason for not taking insurance.
- One participant stated that he did not want to give into terrorists as ground for taking insurance.

Chapter 15

Insurance Provision by Governments and Firms

Since the demand for insurance is influenced by framing, context and the choice set (chapter 14), notions of "objective" demands and "true" welfare of people need articulation in terms of values and forms of evidence, Putnam (2002). Ie insurance analysis and policy advice cannot be mechanical and value free. We need to recognize the scope for:

1 benevolent leadership whereby the government or private insurers manipulate perceptions or the situation facing insurees in ways that aid the welfare of the insured; and

2 malevolent leadership whereby governments or private insurers generate pathologically high levels of demand for insurance.

A person's demand for insurance is less likely to be pathologically high - more likely to indicate where provision of insurance demanded would aid his welfare - in times of normal levels of emotional tension. With the partial exception of session 3 (held in the wake of the Madrid central station bombing and with an overtone of terror in the pre-experiment advertising), in our scenarios there was not an abnormally heightened tension level. We may therefore interpret our participants' decisions and self reports as indicating the following.

When even modest financial sums are at stake, insurance provision at very high premia (very "unfair" rates) improves the welfare of a minority. It does so by reducing their concern about asset loss (negative primary satisfactions) and their worries about asset loss (negative secondary satisfactions). Where the amounts at stake are higher (or people are poorer or the risks higher), being uninsured interferes with normal abilities to plan and commit. In these situations our participants' self reports indicate that the negative secondary satisfactions arising from planning inefficiencies and inabilities to commit, mean that insurance provision even at very high premia (very "unfair" rates) would benefit most people.

In summary, insurance provision has primary and especially secondary satisfactions benefits for citizens. In turn this means that governments may be able to enhance the welfare of their citizens via insurance assistance. It may assist by regulations enhancing the quality of private provision, and by filling gaps where private insurance is unavailable.

The economics of private insurance provision is little understood. It is often modeled as if it could be compatible with competitive markets and with EUT. In fact, insurance provision, is incompatible with both perfect competition and EUT. It comprises exclusively interstate costs, excluded under perfect competition, Pope (forthcoming). Insurance, where genuine (non-fraudulent), reduces downside dispersion of the insured, Pope (forthcoming). But concern about reducing variance - or any other higher moment or measure of downside dispersion - is incompatible with the firm adhering to EUT, Schneeweiß (1968a,b, 1973a,b), Borch (1969), Feldstein (1969).

Firms providing genuine insurance therefore are not perfectly competitive and do not obey EUT - do not maximize expected (utility of) net profits. Instead they employ heuristics or constrained expected profits maximization. To render their insurance provision genuine, their heuristics or constraints must be such as to keep their risk of dishonouring contracts down to the level (implicit) in the insurance contracts.

Private insurance providers are subject to two temptations. One temptation is to oversell insurance benefits. Governments seek to exercise benevolent leadership by limiting this. Governments impose regulations designed to avoid private insurers playing excessively on their potential clients' aspirations for primary satisfactions, such as money, and on their potential clients' desire to avoid negative secondary satisfactions such as worry and being unable to commit and plan. Such regulations include "cooling off periods" before payment of the insurance premium becomes binding, and rules against misleading advertising.

The other temptation of private insurers is to be non-genuine - to offer fraudulent insurance contracts, ie ones for which their reserves are inadequate. In this book's insurance scenario, concern about insurance contracts being dishonoured was not an issue. Our participants had faith in us the experimenters honouring insurance contracts entered. In real life, concern whether insurance contracts are fraudulent, is warranted.

There is a long history of failed private insurance companies - of insurers that did not honour their contracts. For a list of massive multinational insurance companies who have failed in the last 15 years, see Chen (2006). Citizens suffer the negative secondary satisfactions of worry and inefficient planning when they are unsure whether the private insurance company will have reserves to honour its insurance contracts.

To reduce such negative secondary satisfactions governments typically exercise benevolent leadership by imposing reserves requirements on private insurance firms. Such government imposed reserves requirements do not merely pander to irrational negative secondary satisfactions arising out of unwarranted irrational concern of insurees about the solvency of insurers. They cater also for warranted concern arising out of the past high frequency of dishonoured insurance contracts.

But economists trained in EUT cannot recognise such warranted concern, such negative secondary satisfactions arising out of dishonoured contracts. EUT trained economists perceive government imposed reserves requirements as simply undesirable constraints on international insurance service supply. Such EUT policy advisers have "liberalized" conditions for insurance companies. Eg they have had the European Union Commission reduce reserve requirements.

In the aftermath of this deregulation, the governments of France, German and Great Britain have needed to use tax payers' money to rescue their nationally domiciled insurance companies that were deemed "too big to fail", Pope (forthcoming). Thus to alleviate the costs of risk bearing that fraudulent insurance contracts impose on their citizens, governments need to seek policy advisers outside the EUT perspective. EUT's limit to a sub-set of primary satisfactions precludes EUT policy advisers understanding the pertinence of *knowledge ahead*, the pertinence of secondary satisfactions in the form of being able to rely on insurers honouring their contracts. EUT policy advisers' lack understanding:

1 of negative secondary satisfactions from fraud and

2 of how insurers in "liberalized" markets are enticed (even "forced" by competitive pressures) to behave more like EUT maximisers and thus behave fraudulently keeping inadequate reserves.

These failures in the EUT policy lens have been partially responsible for the wave of failed insurance companies operating in the European Union.

Governments at times seek also to exercise benevolent leadership by filling the breach when private insurance firms have found their past reserves inadequate and refuse to continue to accept particular forms of insurance. Insurance provision, our book finds, can benefit people. Thus discontinuance of private insurance can reduce their welfare.

This discontinuance of previously available private insurance has occurred in some countries over the last couple of decades in the case of risks from particular health conditions, adventure sports, natural catastrophes and terrorist attacks. Whether government insurance provision to cover market failures is a provision that increases overall welfare depends partly on its insurance provision costs.[1] In the provision of insurance, the government has risk pooling advantages over many private insurance suppliers in this provision, Samuelson (1964), Vickrey (1964) and Pope (forthcoming). These need to be taken into account in any assessment of the desirablility of governments filling particular insurance gaps.

---

[1]For perspectives on appropriate forms of insurance provision against terrorism, catastrophes and private-for-gain attacks, and whether such contracts should be private, public or a mix, see *eg* Kunreuther (2002), Barker (2003), Lakdawalla & Zanjani (2005), Jaffee & Russell (2006).

Government policies alter citizens' risks in various ways, and thus the extent to which its citizens suffer negative secondary satisfactions from perceiving themselves to be uninsured against risks. Governments exercise their scope to alter these perceptions malevolently, when they fail to reveal environmental, health and other risks associated with their policies, or fan irrational, pathological fears. Governments exercise their scope to alter these perceptions benevolently when they quell pathological excessive worries such as those that have arisen from unhealthy reliving of past terrorist attacks.

The US government might for instance consider achieving this by taxing television stations for repeated showings of terrorist incidents beyond that required for information and informed analysis. Or the US government might consider advertising the comparatively minor nature of *past* terrorism damage relative to other benchmarks of frequencies of bad domestic outcomes. Benchmarks might be non-terrorist domestic insurance property claims, domestic deaths from medical negligence or homicide, or as suggested in Gigerenzer (2004), road deaths. A road's death focus might divert people's "worry quota" into a direction aiding them individually into better decision making - that of safer driving.

Not all the secondary dissatisfactions about possible terrorist attacks in the forms of worry and commitment and planning difficulties are irrational / pathological, ones that a benevolent government should seek to dispel by altering people's perceptions and foci. The issue here is which measures reduce the *actual* likelihood of terrorist attacks at a reasonable cost in terms of money and civil liberties.[2] Assessing which policy mixes actually provide terrorism insurance warrants well-designed empirical investigations. It warrants investigations of the quality advocated and supported by Economists for Peace and Security. In short, terrorism and other forms of warfare are applied branches of insurance that merit more research attention from university departments in economics and the other social sciences than has been the case. The primary negative satisfactions from war damage and the secondary negative satisfactions from fears, worries, and the planning inefficiencies of defensive arms build-ups need to be documented, assessed.

[2]The Bush government has sought to reduce this by heightened national defence expenditure and reduced civil liberties. A complementary approach advocated *eg* in Roberts (2004) and Webster (2004), is to address the injustices as perceived by the terrorists. For perspectives on the causes of terrorism and means of reducing it, see, *eg* Bernholz (2004), Blomberg et al. (2003), Brück & Wickström (2004), Chen & Siems (2004), Fitzgerald (2004), Jain & Mukand (2004), Frey & Luechinger (2004), Nitsch & Schumacher (2004), Foster (2006).

Chapter 16

Summary and Directions for Further Research

Our findings disconfirm two widespread assumptions. One is that favoured in economics of EUT choosers with so-called constant relative risk aversion (involving a concave "as if certain" mapping from outcomes into utilities). The other is that favoured in finance of EUT choosers with so-called risk neutrality (involving a linear "as if certain" mapping from outcomes into utilities).

EUT and other standard rank dependent theories such as CPT, Cumulative Prospect Theory, have here been shown to be undesirably narrow in considering only a sub-set of what are here termed primary satisfactions, namely satisfactions that are independent of *knowledge ahead.* Replacing the inherently static framework of these theories with an evolving *stage of knowledge ahead* framework would include decision makers' reasonable desire to take into account secondary satisfactions pertaining to the emotions and inter-temporal resource allocation both of which depend on the anticipated evolution in knowledge ahead.

Thus it is desirable to move outside the knowledge-wise static framework of EUT and other rank dependent theories. We have indicated how an appreciation of the fundamental role of anticipating a *change in knowledge ahead* can allow scientists to consistently investigate all major sources of satisfactions, secondary as well as primary. We have indicated how to frame and enrich theories and experimental set-ups so as to elicit the information required to develop an understanding of the range of satisfactions that do and should enter the decision procedure for reasonable people. Doing so may open the way for scientists to deepen the applications of these principles in insurance and other contexts. This may then enable pertinent empirical work to identify the quantitative significance of different primary and secondary satisfactions, and the popularity of different algorithmic decision procedure in each particular context. These developments in turn may afford decision scientists insights for better describing and predicting how people chose, and for a better grasp of normative issues and thus of prescribing how people, governments, businesses, and other organizations, should choose.

Appendix A

Experimental Results in Detail

Table A.1: Session 1 23rd May 2003, Participants 1-7

Nr	*Sex*	*Status**	*Insurance decisions* 20%	10%	5%	1%	.2%	.1%	*Reasons in case of insurance* avoid worry	utility shape**	other	*Reasons in case of no insurance* thrills	ultra convex utility	other
1	F	Fac										x	x	The probabilty that there is no payout is very low, it also has to be considered that a person is chosen.
2	M	Fac	x	x	x				x		Risk of 1% can be tolerated			
3	M	MBA											x	The insurance is more expensive than the expected loss at risk neutrality.
4	M	Fac												Too low probabilty of an attack proportional to the price, even at a risk of 20% of an attack.
5	M	Fac												80*0.2=16 < 30. I am not that risk averse.
6	M	MBA										x		Since the insurance costs €30, it is too expensive compared to the maximum expected value of an attack (80*0.2=16) and furthermore a bit of thrill is not bad.
7	M	Fac												Even for the worst-case risk of 20% the expected win (without insurance)= 80*0.8 =0*0.02=64 and 64 > 50. If I knew that I win, I might prefer the €50 fixed sum... it is very unlikely that I am selected at all, I can just as well raise the risk a little.

* Fac: faculty members, MBA: MBA students, CO: Computer Officer in the Department of Statistics and Operations research, a participant amalgamated with faculty into a category Fac+ in other tables.

** utility function is concave or not too convex

x denotes "yes"

Table A.2: Session 1 23rd May 2003, Participants 8-15

Nr	*Sex*	*Status**	*Insurance decisions*						*Reasons in case of insurance*			*Reasons in case of no insurance*		
			20%	*10%*	*5%*	*1%*	*.2%*	*.1%*	*avoid worry*	*utility shape***	*other*	*thrills*	*ultra convex utility*	*other*
8	F	MBA										x		Probabilites are relatively low!
9	F	MBA	x							x				
10†	M	Fac				x				x	In this case it is the same for all cases of probability, so €50 is really better than 0			
11	M	Fac	x	x	x	x			x	x				Probability is small I can take the risk.
12	M	MBA	x	x	x						In order to eliminate the risk above a probability of 5%			
13	M	CO	x						x		20% is a too high risk to me, every 5th money carrier is attacked, at the latest at Friday.			
14	M	MBA										x		I will have luck.
15	F	Fac											x	Insurance is too expensive for €80.

* Fac: faculty members, MBA: MBA students, CO: Computer Officer in the Department of Statistics and Operations research, a participant amalgamated with faculty into a category Fac+ in other tables.

** utility function is concave or not too convex

† Non-monotonic participant

x denotes "yes"

Table A.3: Session 2 11th November 2003*, Participants1-3

Nr	*Sex*	*Insurance decisions*						*Reasons in case of insurance*			*Reasons in case of no insurance*			*Willingness to pay*					
		0.10%	*0.20%*	*1%*	*5%*	*10%*	*20%*	*avoid worry*	*utility shape***	*other*	*thrills*	*ultra convex utility*	*other*	*0.10%*	*0.20%*	*1%*	*5%*	*10%*	*20%*
1	F				x	x	x		x		x		In order to win €80, because the probability of an attack between 0.1% and 1% is substantially lower than for example 20%.	0	0	0	4	8	16
2	M												Expected value at the maximum probability of an attack = 64 > 50 respectively E[damage] = 16 < 30	0.08	0.16	0.8	4	8	16
3	M						x		x		x			0	0	0	5	15	30

* all participants doctoral students
** utility function is concave or not too convex
x denotes "yes"

Table A.4: Session 2 11th November 2003*, Participants 4-6

Nr	*Sex*	*Insurance decisions*						*Reasons in case of insurance*			*Reasons in case of no insurance*			*Willingness to pay*					
		0.10%	*0.20%*	*1%*	*5%*	*10%*	*20%*	*avoid worry*	*utility shape***	*other*	*thrills*	*ultra convex utility*	*other*	*0.10%*	*0.20%*	*1%*	*5%*	*10%*	*20%*
4	M						x			In order to insure me in the highest probability (20%). From my point of view, the others are that low, that an attack is unlikely.			The gain is higher at €80.	0	0	0	5	15	30
5	M										x			0.08	0.16	0.8	4	8	16
6	M												The probability whether I win is 1/18. The maximum probability of an attack is 20%. Aggregated, the probability of winning and attack is too low.	0	0	0.2	1.5	3	10

* all participants doctoral students
** utility function is concave or not tco convex
x denotes "yes"

Table A.5: Session 2 11th November 2003*, Participants 7-11

Nr	*Sex*	*Insurance decisions*						*Reasons in case of insurance*			*Reasons in case of no insurance*			*Willingness to pay*					
		0.10%	*0.20%*	*1%*	*5%*	*10%*	*20%*	*avoid worry*	*utility shape***	*other*	*thrills*	*ultra convex utility*	*other*	*0.10%*	*0.20%*	*1%*	*5%*	*10%*	*20%*
7	M												The probability whether I win is 1/18. The maximum probability of an attack is 20%. Aggregated, the probability of winning and attack is too low.	0	0	0.35	3.5	7	15
8	F					x	x		x				The probability is too low.	0	0	0	0	40	40
9	F						x			So that the highest risk does not have to be beared by me alone.		x		0	10	15	20	25	30
10	F	x	x	x	x	x	x		x					1	2	10	10	20	30
11	F				x	x	x		x	High probability that an attack will happen			Low probability that an attack will happen	0	0	5	10	15	20

* all participants doctoral students

** utility function is concave or not too convex

x denotes "yes"

Table A.6: Session 2 11th November 2003*, Participants 12-18

Nr	*Sex*	*Insurance decisions*						*Reasons in case of insurance*			*Reasons in case of no insurance*			*Willingness to pay*					
		0.10%	*0.20%*	*1%*	*5%*	*10%*	*20%*	*avoid worry*	*utility shape***	*other*	*thrills*	*ultra convex utility*	*other*	*0.10%*	*0.20%*	*1%*	*5%*	*10%*	*20%*
12	F			x	x	x	x		x				Very low probability that an attack will happen	0	0	5	10	20	30
13	M					x	x		x				The probability is too low.	0	0	0	5	10	20
14	F	x	x	x	x	x	x	x	x					0.8	1.6	4	9	10	15
15	M				x	x	x		x		x	x		0	5	5	10	20	30
16	M	x	x	x	x	x	x	x			x			1	1	1.5	7.5	15	30
17	F					x	x		x					0	0	0	0	15	30
18	M				x	x	x	x	x					0	0	0.1	0.5	3	6

* all participants doctoral students
** utility function is concave or not too convex
x denotes "yes"

Table A.7: Session 3 16th March 2004, Decisions, Reasons Insured, Participants 1-10

Nr	*Sex*	*Status**	*Insurance decisions*						*Reasons in case of insurance*			
			20%	*10%*	*5%*	*1%*	*0.20%*	*0.10%*	*avoid worry*	*utility shape***	*still insure if pay ahead*	*other*
1	F	MBA	x								x	The chance of the attack being 20% is high enough to spend $15
2	M	MBA	x									I chose only the 200 days, or 20% chance has it made my EMV only $80 vs. $85 for protection
3	M	MBA	x								x	
4	M	MBA	x								x	Value of insurance should be higher than value of keeping money uninsured
5	M	MBA	x	x	x	x	x					I am unwilling to take more than a 1/1000 risk of losing all the money
6	M	MBA										
7	M	MBA	x	x	x					x	x	
8	M	Fac	x								x	Expected value
9	M	MBA	x	x	x				x		x	
10	M	MBA	x									20% chance of attack. My cost is 15%. So I will pay for the protection

* Fac: faculty members, MBA: MBA student.
** utility function is concave or not too convex.
x denotes "yes"

Table A.8: Session 3 16th March 2004 Decisions, Reasons Insured, Participants 11-21

Nr	*Sex*	*Status**	*Insurance decisions*						*Reasons in case of insurance*			
			20%	*10%*	*5%*	*1%*	*0.20%*	*0.10%*	*avoid worry*	*utility shape***	*still insure if pay ahead*	*other*
11†	M	MBA			x					x	x	
12	M	MBA	x	x						x		
13	M	Fac	x							x	x	
14	M	MBA	x	x	x	x			x			
15	M	MBA	x							x	x	
16	M	MBA	x								x	
17	M	MBA										I didn´t choose to purchase any insurance. As I do not believe the time from last attack to be solid indicator of future attack.
18	M	MBA	x	x						x		
19	M	MBA	x	x	x	x	x	x	x			
20	M	MBA	x	x	x					x	x	
21	M	MBA	x									Playing expected value. Insurance on anything below 150 occurence would not make sense

* Fac: faculty members, MBA: MBA student.
** utility function is concave or not too convex.
† Non-monotonic participant
x denotes "yes"

Table A.9: Session 3 16th March, Decisions, Reasons Insured, Participants 22-32

Nr	*Sex*	*Status**	*Insurance decisions*						*Reasons in case of insurance*			
			20%	*10%*	*5%*	*1%*	*0.20%*	*0.10%*	*avoid worry*	*utility shape***	*still insure if pay ahead*	*other*
22	M	MBA	x	x	x				x	x	x	
23	M	Fac	x	x	x	x				x		I am risk averse. For a danger of 1% or more willing to pay.
24	F	MBA	x	x	x	x	x	x		x	x	
25	M	MBA	x						x		x	The expected value considering the chance of attack
26	M	Fac	x									Expected value
27	M	Fac	x									Purely EMV considerations
28	M	MBA	x									EV of \$85 > EV \$80
29	M	MBA	x	x	x	x					x	The possibility of an attack is too high
30	M	MBA	x	x					x			
31	M	MBA	x								x	The probability of an attack is great enough that the insurance is worth \$ 15 (at 20% chance the cost shout be \$20:\$15 is cheap)
32	M	MBA	x	x	x					x	x	

* Fac: faculty members, MBA: an MBA student.
** utility function is concave or not too convex.
x denotes "yes"

Table A.10: Session 3 16th March, Decisions, Reasons Insured, Participants 33-42

Nr	*Sex*	*Status**	*Insurance decisions*						*Reasons in case of insurance*			
			20%	*10%*	*5%*	*1%*	*0.20%*	*0.10%*	*avoid worry*	*utility shape***	*still insure if pay ahead*	*other*
33	M	MBA	x									There are 20% prob to lose all that money my expected pay out to be below 85 so I picked to pay for 20% probability of losing.
34	M	MBA	x	x	x	x	x			x	x	
35	M	MBA	x							x	x	
36	M	MBA	x	x					x	x	x	
37	M	MBA	x								x	EMV player
38	F	MBA	x	x	x				x	x	x	
39	M	MBA	x								x	Math. decisions. Payout x % likelihood of loss=value of insurance. Only if value of policy exceeds price of the premium I buy.
40	M	MBA	x	x	x					x		
41	M	MBA	x							x		
42	M	MBA	x							x	x	

* Fac: faculty members, MBA: an MBA student.
** utility function is concave or not too convex.
x denotes "yes"

Table A.11: Session 3 16th March, Decisions, Reasons Uninsured, Participants 1-10

Nr	*Reasons in case of no insurance*					
	thrills	*ultra convex utility*	*pre-commit for company*	*commit to repay*	*avoid ridicule blame*	*other*
1			x	x	x	Not high enough chance of attack
2			x	x		EMV considerations it didn´t make sense based on probabilities to buy the protection
3			x	x		
4			x	x		Value proposition
5		x		x	x	
6			x	x		The $100 would be extra, not an reduction of my assets
7			x	x		The likelihood of Attack is low enough that cost of assurance was more than the risk of loss. I suppose I am "self insured".
8		x	x	x	x	
9		x	x	x	x	
10				x		The likelihood of attack is below 15%. My cost is 15% so I will not pay in these circumstance.

x denotes "yes"

Table A.12: Session 3 16th March, Decisions, Reasons Uninsured, Participants 11-21

	Reasons in case of no insurance					
Nr	*thrills*	*ultra convex utility*	*pre-commit for company*	*commit to repay*	*avoid ridicule blame*	*other*
11†		x	x	x		Little difference between \$15 and \$100
12	x		x			
13			x	x		
14						\$15 is lot of money for me, so I am willing to take a small chance to win it.
15			x	x		Why pay for a low probability event?
16			x	x	x	
17			x			The price of insurance seemed high relative to the risk of attack
18						Looking at possibility of attack, there is at most only 20% of possibility to lose
19			x	x	x	\$85 is better than 0
20			x	x		Low chances of attack occurs
21			x	x		EM Value

† Non-monotonic participant
x denotes "yes"

Table A.13: Session 3 16th March, Decisions, Reasons Uninsured, Participants 22-42

	Reasons in case of no insurance					
Nr	*thrills*	*ultra convex utility*	*pre-commit for company*	*commit to repay*	*avoid ridicule blame*	*other*
22			x	x		5% is my personal threshold. From my experience in Israel , I learned not to let have terrorist attacks to intervene my life
23			x	x		0,2% looks to me not worth $15 protecting the $100
24						
25			x	x		The expected value of the "game"
26						Expected value, pay = 85 p = 85% do not = P(100), I-P 15% breakeven
27			x			Again EMV conclusion
28			x	x		
29			x	x		The risk of loosing $100 is very low and not worth the protection
30			x			$ 15 is too high a price for the risk 5% and less
31			x	x		The insurance cost is too dear i.e. $15 the potential loss of the $100
32			x	x		
33			x	x		If I buy insurance, I get only $85 with 1/40 prob. If it wasn't an experiment I'd have paid $15, as nothing goes out of my pocket
34			x	x		Negative outflow
35		x	x	x		
36			x			
37			x			EMV player
38			x	x	x	The chance is very small so I think I don´t need to hedge that risk
39			x	x		Math. decisions. Payout x % likelihood of loss=value of insurance. Only if value of policy exceeds price of the premium I buy.
40	x		x		x	
41		x				
42		x	x	x		

x denotes "yes"

Table A.14: Session 3 16th March 2004, Rank Dependent Weights and Timing of Insurance, Participants 1-10

Nr	*Sex*	*Status**	*more weight the worst***	*explanation of more weight the worst*	*more weight the best***	*explanation of more weight the best*	*still insure if pay ahead***	*explanation of still insure if pay ahead*
1	F	MBA	N		N		Y	
2	M	MBA	N	No, I went with probabilities.	N		Y	It wouldn't have changed my selection of only the 200 times in the last 1000 days, as the EMV would actually be lower with the sure payment described here than with the original scenario.
3	M	MBA	N		N		Y	
4	M	MBA	N		N		Y	
5	M	MBA	N	I was unwilling to take more than 1/1000 risk of losing \$100/\$85 → both seemed not to be different	N		N	
6	M	MBA		Did not pay for protection.	N		N	
7	M	MBA	Y	To be honest it is not entirely rational, but I felt comfortable with <1% risk exposure.	N		Y	
8	M	Fac	N	No, but I am not sure I worked it out correctly.	N	No, time & exposure	Y	
9	M	MBA	N		N		Y	
10	M	MBA	N		N	No, would have time weighted the reward.	N	

* Fac: faculty members, MBA: MBA student.
** Y: yes, N: no.

Table A.15: Session 3 16th March 2004, Rank Dependent Weights and Timing of Insurance, Participants 11-20

Nr	*Sex*	*Status**	*more weight the worst***	*explanation of more weight the worst*	*more weight the best***	*explanation of more weight the best*	*still insure if pay ahead***	*explanation of still insure if pay ahead*
11	M	MBA	N		N		Y	
12	M	MBA	N		N		N	
13	M	Fac	N		N		Y	
14	M	MBA	Y	I'm risk averse. → $100 is a lot of money for me.	N	$100 is a lot, but $15 isn't peanuts. My treshold was reached rather unscientifically. I pretty much divided the true percentage by 10 to signify my risk adverse nature, rounded up, and used that as my cutoff.	N	
15	M	MBA	N		Y	20% - "high" probability of attack, 80% - "low" probability of attack (less than 200 days).	Y	
16	M	MBA	N				Y	
17	M	MBA		I did not purchase any insurance in set A. I did not feel that the protection was a worthwile use of funds.		I was more or less just interested in the probability of attack.		I did not make any insurance in set a.
18	M	MBA	N				N	
19	M	MBA	Y	Yes, choosing protection across spectrum.			N	
20	M	MBA	Y	The greater chance of attacks, I pay.	Y	The greater the risk, the more motivated me to pay.	Y	

* Fac: faculty members, MBA: MBA student.
** Y: yes, N: no.

Table A.16: Session 3 16th March 2004, Rank Dependent Weights and Timing of Insurance, Participants 21-30

Nr	*Sex*	*Status**	*more weight the worst***	*explanation of more weight the worst*	*more weight the best***	*explanation of more weight the best*	*still insure if pay ahead***	*explanation of still insure if pay ahead*
21	M	MBA	N		N		N	No, only want insurance if I pay when I receive EMV bellow 85.
22	M	MBA	Y	1. Can I bear the results of the worst case scenario? 2. How greedy I am? 3. My current financial situation.	Y	Yes, it was a matter of greedness. Good outcome - earned the whole $100. Bad outcome - came with nothing, left with nothing.	Y	
23	M	Fac		I did not have the time to try and evaluate my NM utility for money.		As I said, I did not have time to try and estimate my utility for money.		The question is not clear. If I get for sure $100 then, of course I will pay $15.00 (instead of paying nothing and getting nothing).
24	F	MBA	Y	any $ better than nothing, avoid regret, benefits > extra $15			Y	
25	M	MBA	N		N		Y	
26	M	Fac	Y	Probability of attack 15%	Y	Probability of attack 15%	N	
27	M	Fac	N		N		N	
28	M	MBA	N		N		N	
29	M	MBA	N		N		Y	
30	M	MBA	N		N		N	

* Fac: faculty members, MBA: MBA student.
** Y: yes, N: no.

Table A.17: Session 3 16th March 2004, Rank Dependent Weights and Timing of Insurance, Participants 31-42

Nr	*Sex*	*Status**	*more weight the worst***	*explanation of more weight the worst*	*more weight the best***	*explanation of more weight the best*	*still insure if pay ahead***	*explanation of still insure if pay ahead*
31	M	MBA	N		N		Y	
32	M	MBA	N		N		Y	
33	M	MBA	N	I did only probability calculation.	N	Probability calculation.	N	Then all about 40 people so my rough probability of winning is 1/40. That makes my payout my then 15 so I won't buy anything.
34	M	MBA	N		Y		Y	$15 \cdot 5 = 75$ Rec. 85 hence 10 $15 \cdot 6 = 90$ Rec 85 hence -5
35	M	MBA	N	No, although I considered some "other factors" in my determination of non-payments.	N		Y	It would have been more likely if I had the assets.
36	M	MBA	N	I did it by the frequency of attacks and the probability.		Simple probability per days of attack.	Y	
37	M	MBA	N		N		Y	I'm not sure I understand the question, but I would certainly pay $15 to receive $100 guaranteed late. I'm indifferent to timing of 90 minutes.
38	F	MBA	Y	Whether the chance is big enough to make a hedge of risk necessary.	N		Y	Not different. I'd hold my decision as before.
39	M	MBA	N		N		Y	
40	M	MBA	N		N		N	
41	M	MBA	N		N		N	
42	M	MBA	Y	$85 free money > $0.	N		Y	

* Fac: faculty members, MBA: MBA student.
** Y: yes, N: no.

Table A.18: Session 4 28th April 2004*, Participants 1-11

Nr	Sex	Insurance decisions						Reasons in case of insurance		Reasons in case of no insurance			Other secondary satis-factions***			Willingness to pay					
		0.10%	*0.20%*	*1%*	*5%*	*10%*	*20%*	*avoid worry*	*utility shape***	*thrills*	*ultra convex utility*	*other*	*IV*	*V*	*VI*	*0.10%*	*0.20%*	*1%*	*5%*	*10%*	*20%*
1	M									x		Risk loving				0	0	0	0	0	0
2	F	x	x	x	x	x	x		x				x	x	x						
3	M									x			x	x		0	0	0	0	1	2
4	F					x	x		x	x			x	x		1	2	3	6		
5	M						x	x	x		x		x	x	x	0	1	2	3	5	12
6	M											Highest probabilty for 0 (from the dice) = 1/6*1/5				0	0	0	1	2	4
7	F											The highest probability of 20% is still too low for me. Therefore I want to take all or nothing	x	x		1	1.5	2	4	6	8
8	M						x		x	x			x	x		0	0	1	2	5	
9	F	x	x	x	x	x	x		x				x	x	x						
10	F											Probability for an attack is too small			x	0	0	0	0	0	10
11	F										x	Probability is too small	x			0	0	0	0	5	5

* All participants undergraduate students.
** utility function is concave or not too convex
*** IV: pre-commit for company, V: pre-commit to eat, VI: avoid ridicule blame

Table A.19: Session 4 28th April 2004*, Participants 12-20

Nr	*Sex*	Insurance decisions						Reasons in case of insurance		Reasons in case of no insurance			*Other secondary satis-factions****			*Willingness to pay*					
		0.10%	*0.20%*	*1%*	*5%*	*10%*	*20%*	*avoid worry*	*utility shape***	*thrills*	*ultra convex utility*	*other*	*IV*	*V*	*VI*	*0.10%*	*0.20%*	*1%*	*5%*	*10%*	*20%*
12	F	x	x	x	x	x	x		x				x	x							
13	M					x	x		x	x			x	x		1	1	5	8		
14	F				x	x	x		x	x				x		0	0	5			
15	F											For me personally, the risk that an attack takes place is too low. Above a risk of about 40% I probably had decided for insurance.		x		0	0	1	3	5	8
16	M											Because the probability that an attack takes place is low	x	x	x						
17	M	x	x	x	x	x	x		x				x	x	x						
18	F									x		Ready to take risk; all or nothing	x	x		0	0	0	3	5	5
19	M										x		x	x		0	0	0	0	0	12
20†	F	x														12	0	3	4	5	6

* All participants undergraduate students.
** utility function is concave or not too convex
*** IV: pre-commit for company, V: pre-commit to eat, VI: avoid ridicule blame
† Non-monotonic participant

Table A.20: Session 5 28th April 2004*, Participants 1-4

Nr	*Sex*	Insurance decisions *20%*	*10%*	*5%*	*1%*	*0.20%*	*0.10%*	Reasons in case of insurance *utility shape***	*other*	Reasons in case of no insurance *thrills*	*ultra* convex *utility*	*other*	*Other secondary satis-factions**** *IV*	*V*	*VI*	*Willingness to pay* *20%*	*10%*	*5%*	*1%*	*0.20%*	*0.10%*
1	M									x		Nothing to lose	x	x		12	8	2	1	1	1
2	M	x	x						1/20 probability that I am chosen is rather small, if a 6 or 5 is drawn from the dice, than the probability of winning nothing is high.			Probability is so small	x	x	x			5	1	0	0
3	F									x						17	15	5	2	2	2
4	M	x							20% is relatively large probability. In this case, I am satisfied with €68.			In the other cases, the probability of an attack is that small that €12 is too expensive.	x				8	4	0.5	0.1	0.1

* All participants undergraduate students.
** utility function is concave or not too convex
*** IV: pre-commit for company, V: pre-commit to eat, VI: avoid ridicule blame
x denotes ”yes”

Table A.21: Session 5 28th April 2004*, Participants 5-7

Nr	*Sex*	Insurance decisions *20%*	*10%*	*5%*	*1%*	*0.20%*	*0.10%*	Reasons in case of insurance *utility shape***	*other*	Reasons in case of no insurance *thrills*	*ultra* convex *utility*	*other*	*Other secondary satis-factions**** *IV*	*V*	*VI*	*Willingness to pay* *20%*	*10%*	*5%*	*1%*	*0.20%*	*0.10%*
5	M											All or nothing. I consider the given chance as an opportunity to gamble.				5	4	3	2	0	0
6	M											The probability of an attack is too small to take the risk of losing € 12	x	x		5	2.5	1.25	0.5	0.1	0.05
7	F	x	x	x	x	x	x		I pay € 12 once and I am insured for all attacks				x	x							

* All participants undergraduate students.
** utility function is concave or not too convex
*** IV: pre-commit for company, V: pre-commit to eat, VI: avoid ridicule blame
x denotes "yes"

Table A.22: Session 5 28th April 2004*, Participants 8 and 9

Nr	*Sex*	Insurance decisions *20%*	*10%*	*5%*	*1%*	*0.20%*	*0.10%*	Reasons in case of insurance *utility shape***	*other*	Reasons in case of no insurance *thrills*	*ultra* convex *utility*	*other*	*Other secondary satis-factions**** *IV*	*V*	*VI*	*Willingness to pay* *20%*	*10%*	*5%*	*1%*	*0.20%*	*0.10%*
8	M									x		...and because the occurance of an attack is unlikely and it is not a VERY large amount of money. At a high amount (eg more than € 1000 I would always decide for insurance)	x			8	5	3	2	1	0.5
9	M											With nothing at stake and without risk (80-12) get the maximum.		x	x	10	7	5	3	1	1

* All participants undergraduate students.
** utility function is concave or not too convex
*** IV: pre-commit for company, V: pre-commit to eat, VI: avoid ridicule blame
x denotes "yes"

Table A.23: Session 5 28th April 2004*, Participant 10

Nr	*Sex*	Insurance decisions *20%*	*10%*	*5%*	*1%*	*0.20%*	*0.10%*	Reasons in case of insurance *utility shape***	*other*	Reasons in case of no insurance *thrills*	*ultra* convex *utility*	*other*	*Other secondary satis-factions**** *IV*	*V*	*VI*	*Willingness to pay* *20%*	*10%*	*5%*	*1%*	*0.20%*	*0.10%*
10	M	x	x	x	x	x	x		12 is not much for insurance, and so I get money in any case. If 80 or 12 - it does not matter, but if everthing is lost that'd be a pity.				x								

* All participants undergraduate students.
** utility function is concave or not too convex
*** IV: pre-commit for company, V: pre-commit to eat, VI: avoid ridicule blame
x denotes "yes"

Table A.24: Session 5 28th April 2004*, Participants 11 and 12

Nr	*Sex*	Insurance decisions *20%*	*10%*	*5%*	*1%*	*0.20%*	*0.10%*	Reasons in case of insurance *utility shape***	*other*	Reasons in case of no insurance *thrills*	*ultra convex utility*	*other*	*Other secondary satisfactions**** *IV*	*V*	*VI*	*Willingness to pay* *20%*	*10%*	*5%*	*1%*	*0.20%*	*0.10%*
11	M											I had nothing at the beginning of the experiment. Thus I can only win something or leave with as much as I had before.	x			8	5	0	0	0	0
12	F	x	x					x				I have nothing to lose, because right now I have 0. Actually I can only win, so why not risk a little bit. If I do not get something at lower than 5%, than it was not supposed to happen.	x	x				6	2	0	0

* All participants undergraduate students.
** utility function is concave or not too convex
*** IV: pre-commit for company, V: pre-commit to eat, VI: avoid ridicule blame
† Non-monotonic participant
x denotes "yes"

Table A.25: Session 5 28th April 2004*, Participants 13-18

Nr	*Sex*	Insurance decisions *20%*	*10%*	*5%*	*1%*	*0.20%*	*0.10%*	Reasons in case of insurance *utility shape***	*other*	Reasons in case of no insurance *thrills*	*ultra* convex utility	*other*	*Other secondary satisfactions**** *IV*	*V*	*VI*	*Willingness to pay* *20%*	*10%*	*5%*	*1%*	*0.20%*	*0.10%*
13	M	x						x				Is not that likely	x	x	x		7	2	1	0.2	0.1
14	F	x	x					x			x	In all other cases the probability is too small for an insurance	x	x				5	0	0	0
15†	F			x	x	x	x				x	The probability that an attack occurs is low at 20% and 10% ⇒ just rational				2	5				
16†	M		x	x	x					x	x	Probability of an attack is very low	x	x		30	10	0	0	0	0
17†	F			x	x		x	x					x	x	x	10	5			0.5	
18†	M				x			x					x		x	20	17	15		10	5

* All participants undergraduate students.
** utility function is concave or not too convex
*** IV: pre-commit for company, V: pre-commit to eat, VI: avoid ridicule blame
† Non-monotonic participant
x denotes "yes"

Appendix B

Utility Shapes

Figure 1: "Utilities" deduced from EUT for participant 1 in session 2.

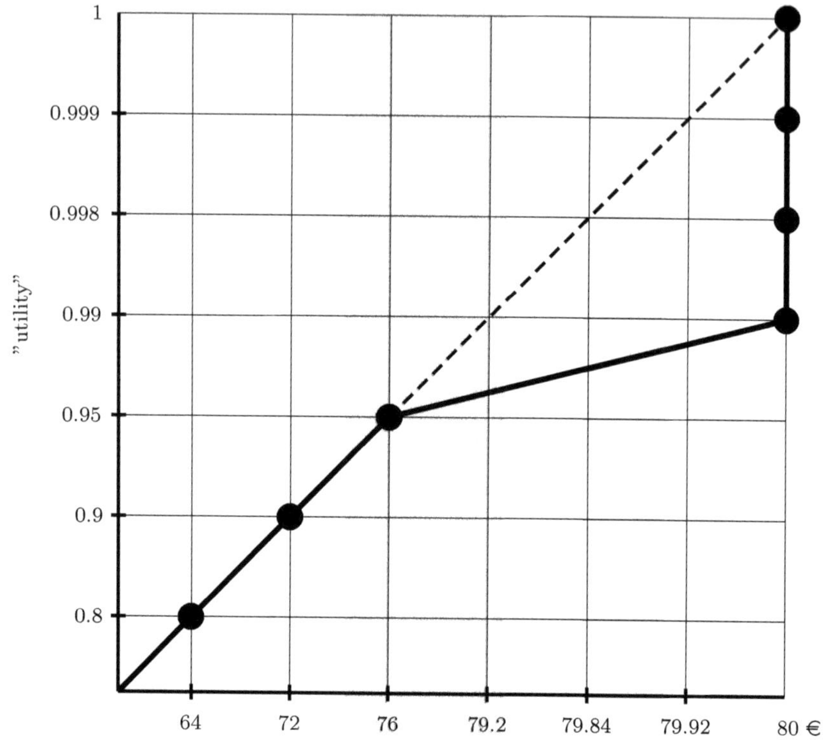

- - - - linear NM utility or fair bet price benchmark
——— participant 1 in session 2's EUT imputed "utilities" with linear interpolation between each imputed "utility".

Risk of attack	**0.2**	**0.1**	**0.05**	**0.01**	**0.002**	**0.001**
x denotes chose to insure	x	x	x			
WTP in €	16	8	4	0	0	0
Compared to linear NM utility or fair bet	=	=	=	<	<	<
Certainty equivalent: 80 - WTP	64	72	76	80	80	80
Classification	Linear then Concave then Convex-Vertical Inconsistency of WTP and insurance decisions					

1 The scales on the axes are non-linear, but chosen such that (subject to graph grid limits), the fair bet choice benchmark denoting linear utility under EUT, is linear

2 The vertical axis shows the probability of obtaining €80 without insurance. The thick black line shows the connection of this probablity with a certainty equivalent amount, namely €80 minus the willingness to pay for insurance at this probability, with linear interpolations between adjacent amounts (circles). Under the assumption that the participant obeys the axioms of expected utility theory, one can normalize the utilities of €0 and €80 as 0 and 1 and the vertical axis represents the utilities and the thick black line denotes the utility function of the participant. But since the thick black line contains a vertical section, the participant does not obey the axioms of expected utility theory.

Figure 2: "Utilities" deduced from EUT for participant 2 in session 2.

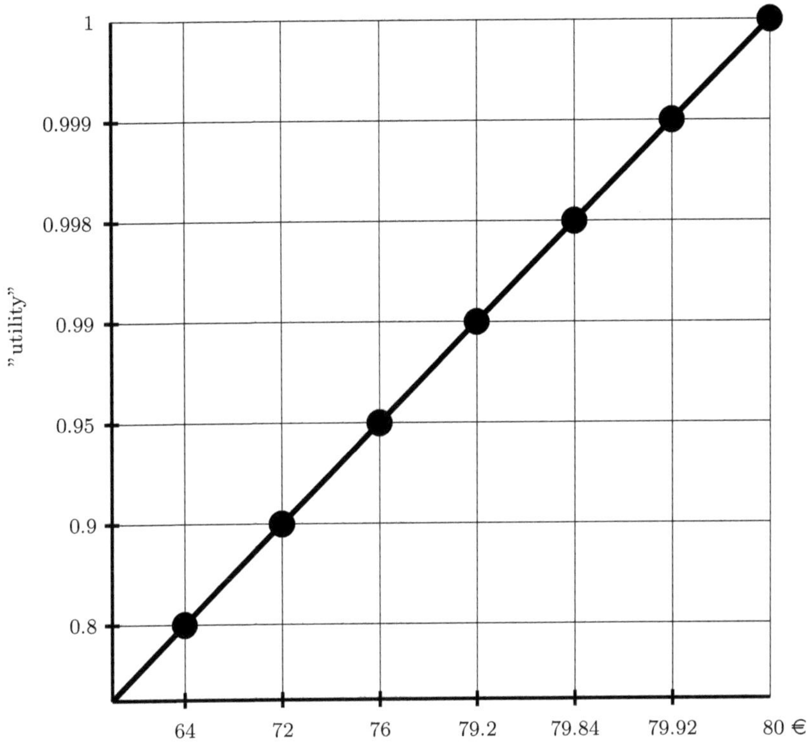

- - - - - linear NM utility or fair bet price benchmark
——— participant 2 in session 2's EUT imputed "utilities" with linear interpolation between each imputed "utility".

Risk of attack	0.2	0.1	0.05	0.01	0.002	0.001
x denotes chose to insure						
WTP in €	16	8	4	0.8	0.16	0.08
Compared to linear NM utility or fair bet	=	=	=	=	=	=
Certainty equivalent: 80 - WTP	64	72	76	79.2	79.84	79.92
Classification	Linear Consistency of WTP and insurance decisions					

1 The scales on the axes are non-linear, but chosen such that (subject to graph grid limits), the fair bet choice benchmark denoting linear utility under EUT, is linear

2 The vertical axis shows the probability of obtaining €80 without insurance. The thick black line shows the connection of this probablity with a certainty equivalent amount, namely €80 minus the willingness to pay for insurance at this probability, with linear interpolations between adjacent amounts (circles). Under the assumption that the participant obeys the axioms of expected utility theory, one can normalize the utilities of €0 and €80 as 0 and 1 and the vertical axis represents the utilities and the thick black line denotes the utility function of the participant.

Figure 3: "Utilities" deduced from EUT for participant 3 in session 2.

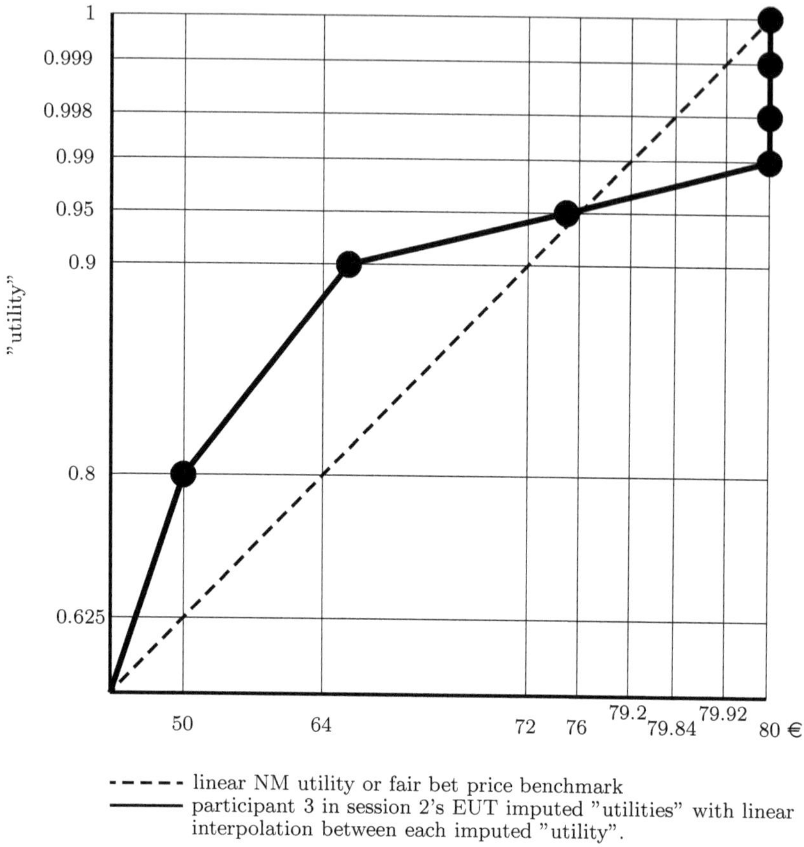

Risk of attack	0.2	0.1	0.05	0.01	0.002	0.001
x denotes chose to insure	x					
WTP in €	30	15	5	0	0	0
Compared to linear NM utility or fair bet	>	>	>	<	<	<
Certainty equivalent: 80 - WTP	50	65	75	80	80	80
Classification	Concave then Convex-Vertical Consistency of WTP and insurance decisions					

1 The scales on the axes are non-linear, but chosen such that (subject to graph grid limits), the fair bet choice benchmark denoting linear utility under EUT, is linear

2 The vertical axis shows the probability of obtaining €80 without insurance. The thick black line shows the connection of this probablity with a certainty equivalent amount, namely €80 minus the willingness to pay for insurance at this probability, with linear interpolations between adjacent amounts (circles). Under the assumption that the participant obeys the axioms of expected utility theory, one can normalize the utilities of €0 and €80 as 0 and 1 and the vertical axis represents the utilities and the thick black line denotes the utility function of the participant. But since the thick black line contains a vertical section, the participant does not obey the axioms of expected utility theory.

Figure 4: "Utilities" deduced from EUT for participant 4 in session 2.

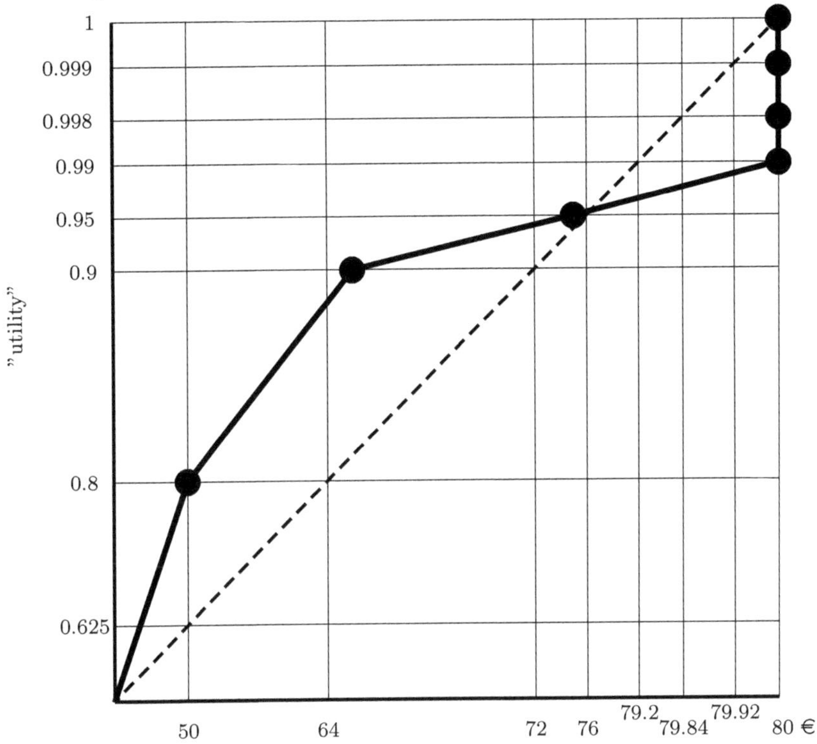

- - - - - linear NM utility or fair bet price benchmark
——— participant 4 in session 2's EUT imputed "utilities" with linear interpolation between each imputed "utility".

Risk of attack	**0.2**	**0.1**	**0.05**	**0.01**	**0.002**	**0.001**
x donotes chose to insure	x					
WTP in €	30	15	5	0	0	0
Compared to linear NM utility or fair bet	>	>	>	<	<	<
Certainty equivalent: 80 - WTP	50	65	75	80	80	80
Classification	Concave then Convex-Vertical Consistency of WTP and insurance decisions					

1 The scales on the axes are non-linear, but chosen such that (subject to graph grid limits), the fair bet choice benchmark denoting linear utility under EUT, is linear

2 The vertical axis shows the probability of obtaining €80 without insurance. The thick black line shows the connection of this probablity with a certainty equivalent amount, namely €80 minus the willingness to pay for insurance at this probability, with linear interpolations between adjacent amounts (circles). Under the assumption that the participant obeys the axioms of expected utility theory, one can normalize the utilities of €0 and €80 as 0 and 1 and the vertical axis represents the utilities and the thick black line denotes the utility function of the participant. But since the thick black line contains a vertical section, the participant does not obey the axioms of expected utility theory.

Figure 5: "Utilities" deduced from EUT for participant 5 in session 2.

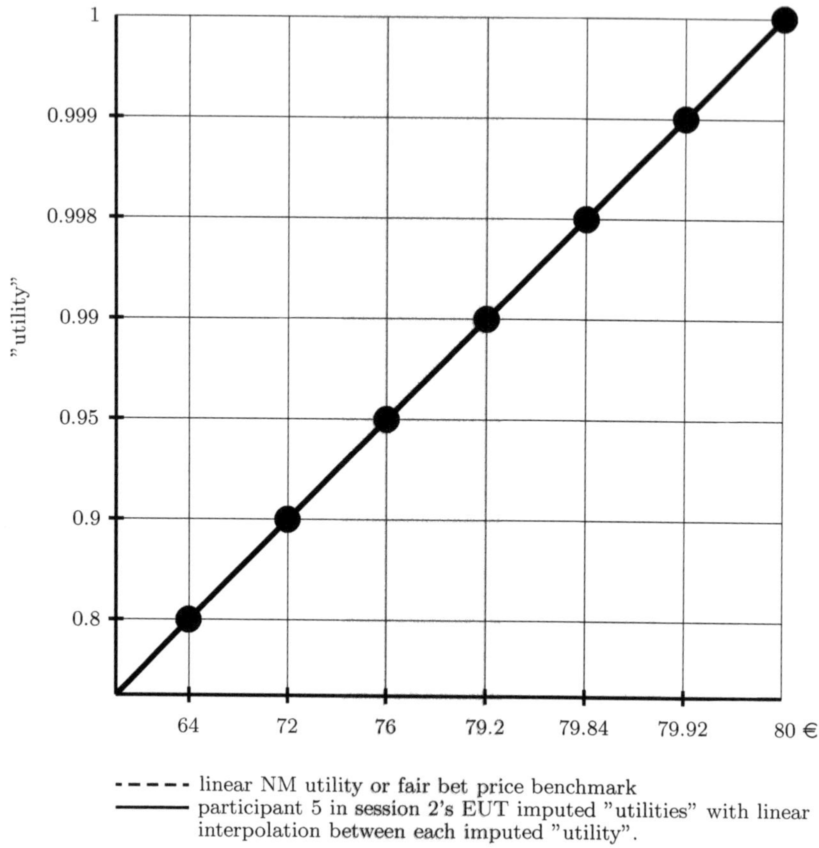

Risk of attack	0.2	0.1	0.05	0.01	0.002	0.001
x denotes chose to insure						
WTP in €	16	8	4	0.8	0.16	0.08
Compared to linear NM utility or fair bet	=	=	=	=	=	=
Certainty equivalent: 80 - WTP	64	72	76	79.2	79.84	79.92
Classification	Linear Consistency of WTP and insurance decisions					

1 The scales on the axes are non-linear, but chosen such that (subject to graph grid limits), the fair bet choice benchmark denoting linear utility under EUT, is linear

2 The vertical axis shows the probability of obtaining €80 without insurance. The thick black line shows the connection of this probablity with a certainty equivalent amount, namely €80 minus the willingness to pay for insurance at this probability, with linear interpolations between adjacent amounts (circles). Under the assumption that the participant obeys the axioms of expected utility theory, one can normalize the utilities of €0 and €80 as 0 and 1 and the vertical axis represents the utilities and the thick black line denotes the utility function of the participant.

Figure 6: "Utilities" deduced from EUT for participant 6 in session 2.

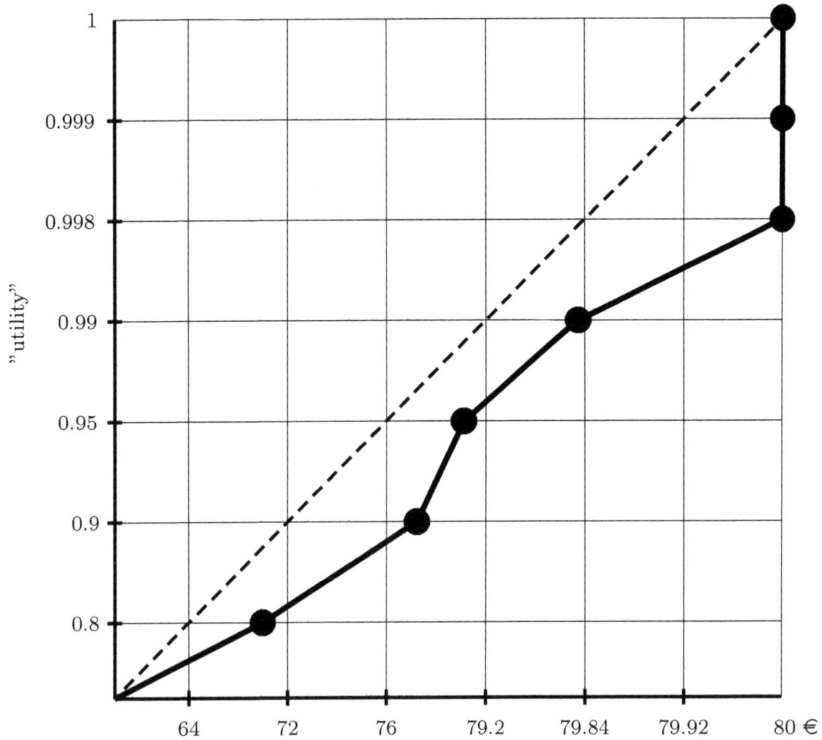

- - - - - linear NM utility or fair bet price benchmark
——— participant 6 in session 2's EUT imputed "utilities" with linear interpolation between each imputed "utility".

Risk of attack	0.2	0.1	0.05	0.01	0.002	0.001
x denotes chose to insure						
WTP in €	10	3	1.5	0.2	0	0
Compared to linear NM utility or fair bet	<	<	<	<	<	<
Certainty equivalent: 80 - WTP	70	77	78.5	79.8	80	80
Classification	Convex then Concave then Convex-Vertical Consistency of WTP and insurance decisions					

1 The scales on the axes are non-linear, but chosen such that (subject to graph grid limits), the fair bet choice benchmark denoting linear utility under EUT, is linear

2 The vertical axis shows the probability of obtaining €80 without insurance. The thick black line shows the connection of this probablity with a certainty equivalent amount, namely €80 minus the willingness to pay for insurance at this probability, with linear interpolations between adjacent amounts (circles). Under the assumption that the participant obeys the axioms of expected utility theory, one can normalize the utilities of €0 and €80 as 0 and 1 and the vertical axis represents the utilities and the thick black line denotes the utility function of the participant. But since the thick black line contains a vertical section, the participant does not obey the axioms of expected utility theory.

Figure 7: "Utilities" deduced from EUT for participant 7 in session 2.

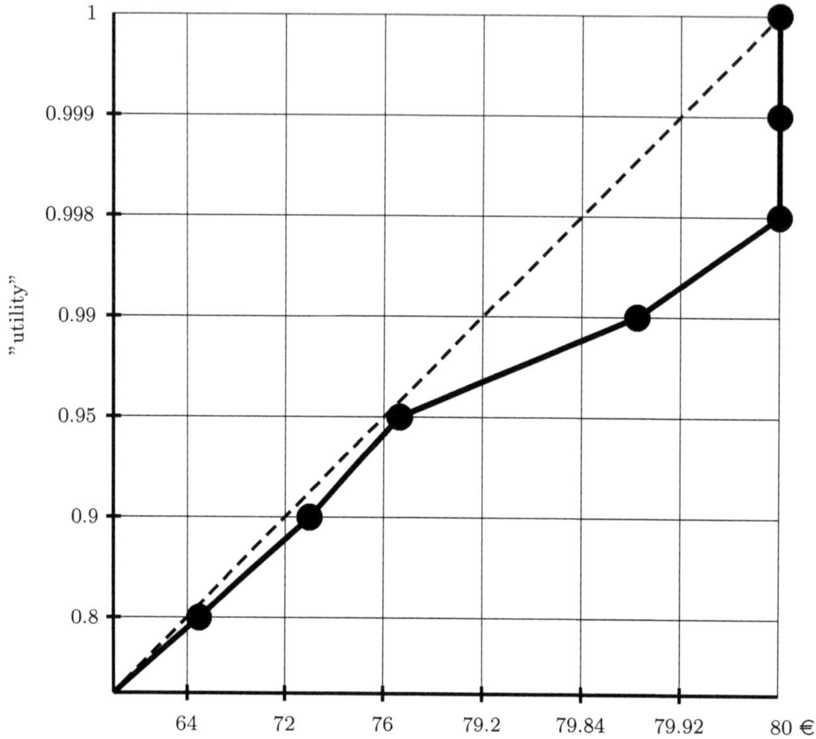

- - - - - linear NM utility or fair bet price benchmark
——— participant 7 in session 2's EUT imputed "utilities" with linear interpolation between each imputed "utility".

Risk of attack	0.2	0.1	0.05	0.01	0.002	0.001
x denotes chose to insure						
WTP in €	15	7	3.5	0.35	0	0
Compared to linear NM utility or fair bet	<	<	<	<	<	<
Certainty equivalent: 80 - WTP	65	73	76.5	79.65	80	80
Classification	Convex then Concave then Convex-Vertical Consistency of WTP and insurance decisions					

1 The scales on the axes are non-linear, but chosen such that (subject to graph grid limits), the fair bet choice benchmark denoting linear utility under EUT, is linear

2 The vertical axis shows the probability of obtaining €80 without insurance. The thick black line shows the connection of this probablity with a certainty equivalent amount, namely €80 minus the willingness to pay for insurance at this probability, with linear interpolations between adjacent amounts (circles). Under the assumption that the participant obeys the axioms of expected utility theory, one can normalize the utilities of €0 and €80 as 0 and 1 and the vertical axis represents the utilities and the thick black line denotes the utility function of the participant. But since the thick black line contains a vertical section, the participant does not obey the axioms of expected utility theory.

Figure 8: "Utilities" deduced from EUT for participant 8 in session 2.

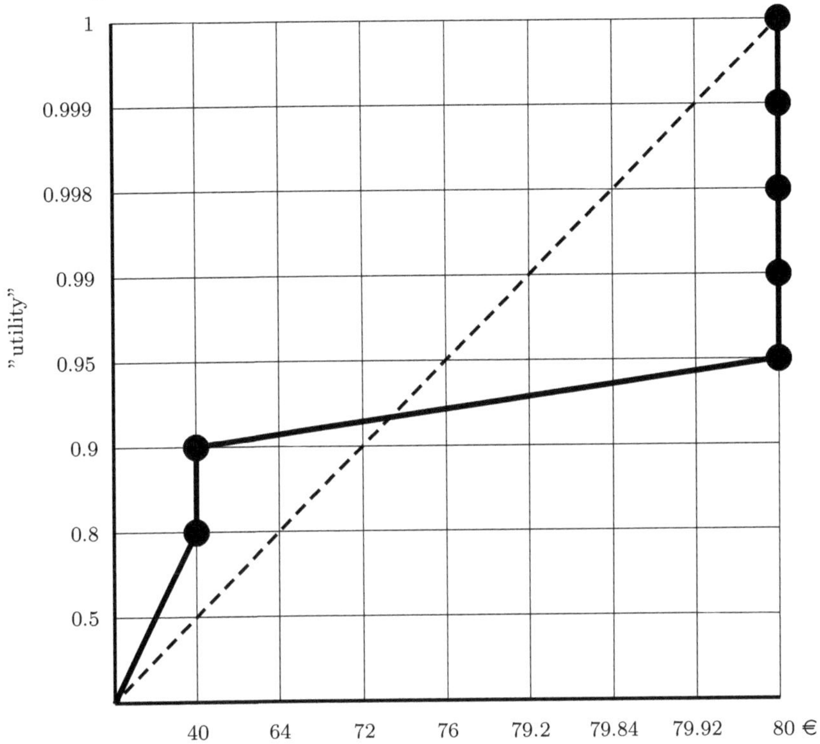

Risk of attack	0.2	0.1	0.05	0.01	0.002	0.001
x denotes chose to insure	x	x				
WTP in €	40	40	0	0	0	0
Compared to linear NM utility or fair bet	>	>	<	<	<	<
Certainty equivalent: 80 - WTP	40	40	80	80	80	80
Classification	Convex-Vertical then Concave then Convex-Vertical Consistency of WTP and insurance decisions					

1 The scales on the axes are non-linear, but chosen such that (subject to graph grid limits), the fair bet choice benchmark denoting linear utility under EUT, is linear

2 The vertical axis shows the probability of obtaining €80 without insurance. The thick black line shows the connection of this probablity with a certainty equivalent amount, namely €80 minus the willingness to pay for insurance at this probability, with linear interpolations between adjacent amounts (circles). Under the assumption that the participant obeys the axioms of expected utility theory, one can normalize the utilities of €0 and €80 as 0 and 1 and the vertical axis represents the utilities and the thick black line denotes the utility function of the participant. But since the thick black line contains vertical sections, the participant does not obey the axioms of expected utility theory.

Figure 9: "Utilities" deduced from EUT for participant 9 in session 2.

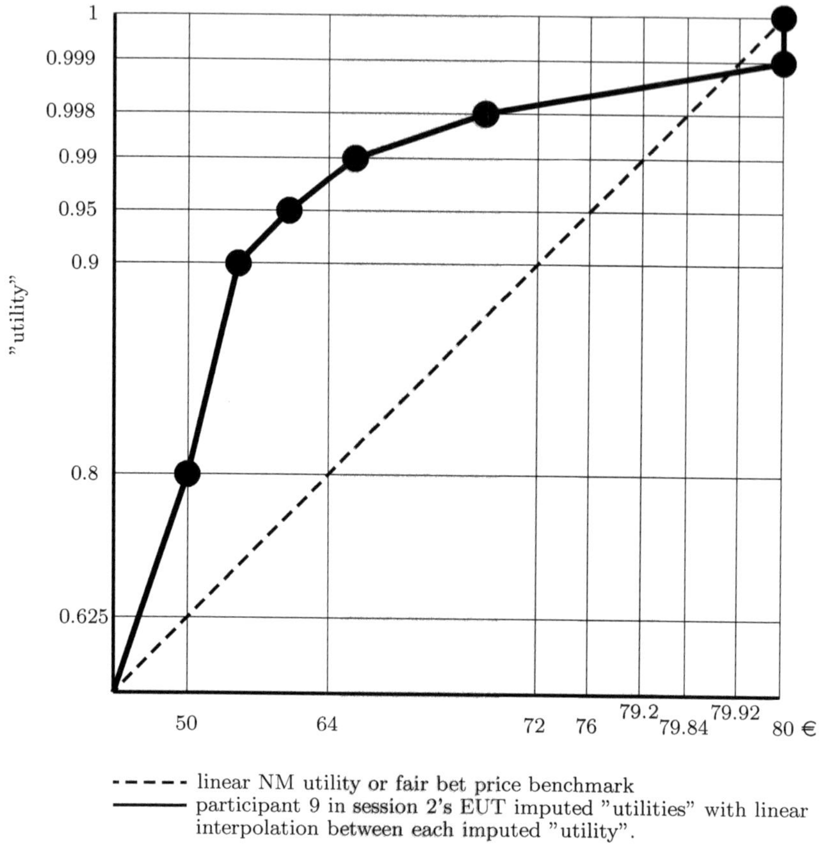

Risk of attack	0.2	0.1	0.05	0.01	0.002	0.001
x denotes chose to insure	x					
WTP in €	30	25	20	15	10	0
Compared to linear NM utility or fair bet	>	>	>	>	>	<
Certainty equivalent: 80 - WTP	50	55	60	65	70	80
Classification	Convex then Concave then Convex-Vertical Consistency of WTP and insurance decisions					

1 The scales on the axes are non-linear, but chosen such that (subject to graph grid limits), the fair bet choice benchmark denoting linear utility under EUT, is linear

2 The vertical axis shows the probability of obtaining €80 without insurance. The thick black line shows the connection of this probablity with a certainty equivalent amount, namely €80 minus the willingness to pay for insurance at this probability, with linear interpolations between adjacent amounts (circles). Under the assumption that the participant obeys the axioms of expected utility theory, one can normalize the utilities of €0 and €80 as 0 and 1 and the vertical axis represents the utilities and the thick black line denotes the utility function of the participant. But since the thick black line contains a vertical section, the participant does not obey the axioms of expected utility theory.

Figure 10: "Utilities" deduced from EUT for participant 10 in session 2.

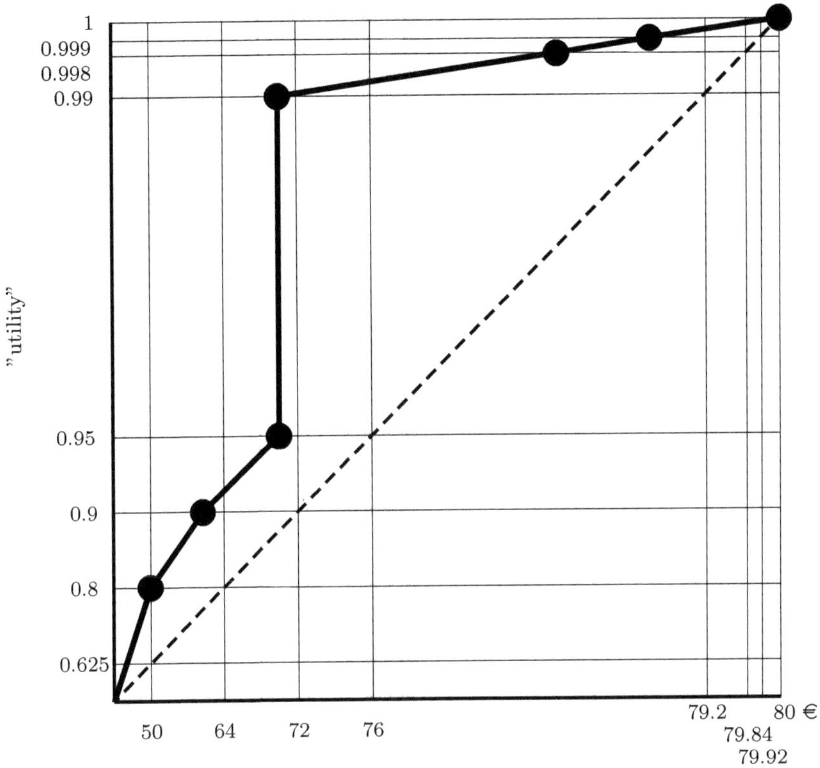

- - - - - linear NM utility or fair bet price benchmark
——— participant 10 in session 2's EUT imputed "utilities" with linear interpolation between each imputed "utility".

Risk of attack	**0.2**	**0.1**	**0.05**	**0.01**	**0.002**	**0.001**
x denotes chose to insure	x	x	x	x	x	x
WTP in €	30	20	10	10	2	1
Compared to linear NM utility or fair bet	>	>	>	>	>	>
Certainty equivalent: 80 - WTP	50	60	70	70	78	79
Classification	Concave then Convex-Vertical then Concave then Linear Inconsistency of WTP and insurance decisions					

1 The scales on the axes are non-linear, but chosen such that (subject to graph grid limits), the fair bet choice benchmark denoting linear utility under EUT, is linear

2 The vertical axis shows the probability of obtaining €80 without insurance. The thick black line shows the connection of this probablity with a certainty equivalent amount, namely €80 minus the willingness to pay for insurance at this probability, with linear interpolations between adjacent amounts (circles). Under the assumption that the participant obeys the axioms of expected utility theory, one can normalize the utilities of €0 and €80 as 0 and 1 and the vertical axis represents the utilities and the thick black line denotes the utility function of the participant. But since the thick black line contains a vertical section, the participant does not obey the axioms of expected utility theory.

Figure 11: "Utilities" deduced from EUT for participant 11 in session 2.

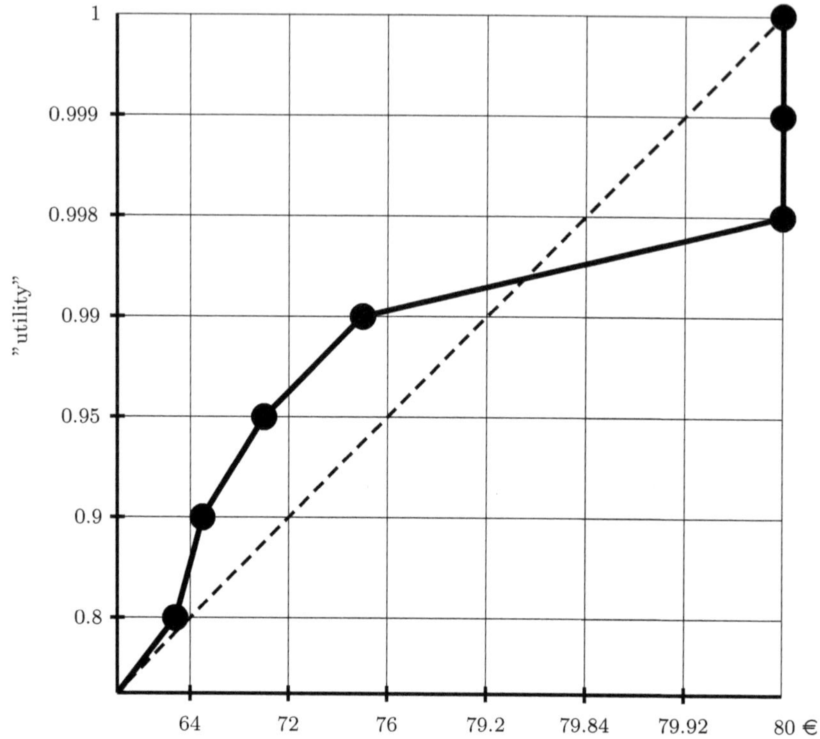

- - - - - linear NM utility or fair bet price benchmark
——— participant 11 in session 2's EUT imputed "utilities" with linear interpolation between each imputed "utility".

Risk of attack	**0.2**	**0.1**	**0.05**	**0.01**	**0.002**	**0.001**
x denotes chose to insure	x	x	x			
WTP in €	20	15	10	5	0	0
Compared to linear NM utility or fair bet	>	>	>	>	<	<
Certainty equivalent: 80 - WTP	60	65	70	75	80	80
Classification	Convex then Concave then Convex-Vertical Inconsistency of WTP and insurance decisions					

1 The scales on the axes are non-linear, but chosen such that (subject to graph grid limits), the fair bet choice benchmark denoting linear utility under EUT, is linear

2 The vertical axis shows the probability of obtaining € 80 without insurance. The thick black line shows the connection of this probablity with a certainty equivalent amount, namely € 80 minus the willingness to pay for insurance at this probability, with linear interpolations between adjacent amounts (circles). Under the assumption that the participant obeys the axioms of expected utility theory, one can normalize the utilities of € 0 and € 80 as 0 and 1 and the vertical axis represents the utilities and the thick black line denotes the utility function of the participant. But since the thick black line contains a vertical section, the participant does not obey the axioms of expected utility theory.

Figure 12: "Utilities" deduced from EUT for participant 12 in session 2.

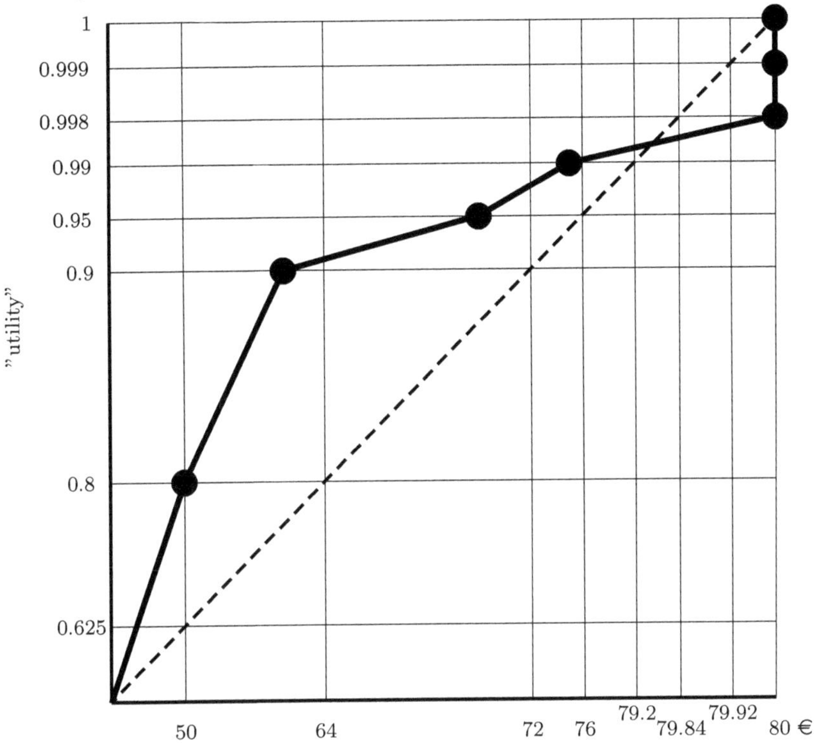

- - - - linear NM utility or fair bet price benchmark
—— participant 12 in session 2's EUT imputed "utilities" with linear interpolation between each imputed "utility".

Risk of attack	0.2	0.1	0.05	0.01	0.002	0.001
x denotes chose to insure	x	x	x	x		
WTP in €	30	20	10	5	0	0
Compared to linear NM utility or fair bet	>	>	>	>	<	<
Certainty equivalent: 80 - WTP	50	60	70	75	80	80
Classification	Concave then Convex then Concave then Convex-Vertical Inconsistency of WTP and insurance decisions					

1 The scales on the axes are non-linear, but chosen such that (subject to graph grid limits), the fair bet choice benchmark denoting linear utility under EUT, is linear

2 The vertical axis shows the probability of obtaining €80 without insurance. The thick black line shows the connection of this probablity with a certainty equivalent amount, namely €80 minus the willingness to pay for insurance at this probability, with linear interpolations between adjacent amounts (circles). Under the assumption that the participant obeys the axioms of expected utility theory, one can normalize the utilities of €0 and €80 as 0 and 1 and the vertical axis represents the utilities and the thick black line denotes the utility function of the participant. But since the thick black line contains a vertical section, the participant does not obey the axioms of expected utility theory.

Figure 13: "Utilities" deduced from EUT for participant 13 in session 2.

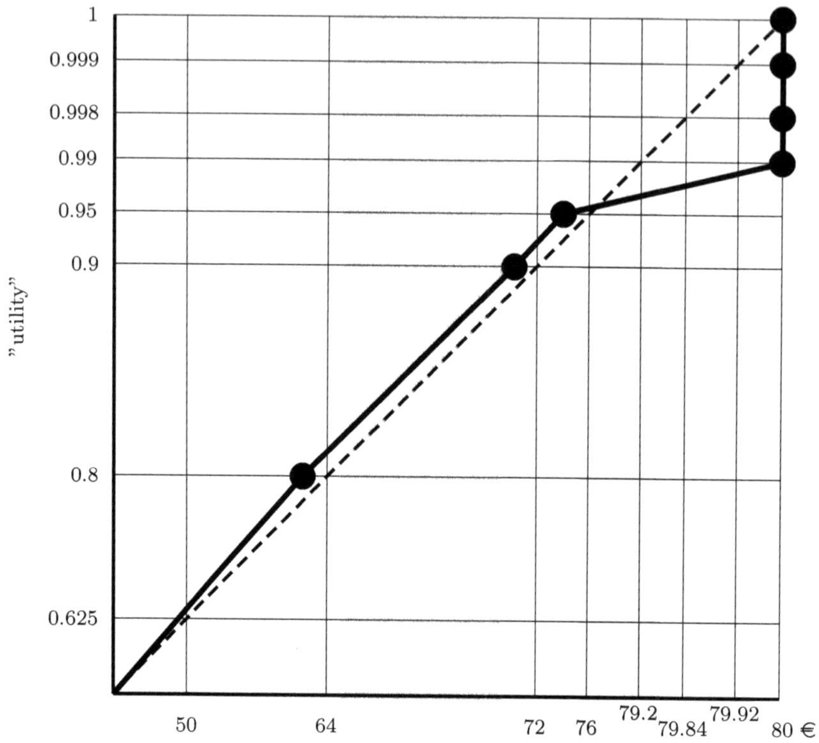

- - - - linear NM utility or fair bet price benchmark
——— participant 13 in session 2's EUT imputed "utilities" with linear interpolation between each imputed "utility".

Risk of attack	0.2	0.1	0.05	0.01	0.002	0.001
x denotes chose to insure	x	x				
WTP in €	20	10	5	0	0	0
Compared to linear NM utility or fair bet	>	>	>	<	<	<
Certainty equivalent: 80 - WTP	60	70	75	80	80	80
Classification	Concave then Linear then Concave then Convex-Vertical. Inconsistency of WTP and insurance decisions					

1 The scales on the axes are non-linear, but chosen such that (subject to graph grid limits), the fair bet choice benchmark denoting linear utility under EUT, is linear
2 The vertical axis shows the probability of obtaining € 80 without insurance. The thick black line shows the connection of this probablity with a certainty equivalent amount, namely € 80 minus the willingness to pay for insurance at this probability, with linear interpolations between adjacent amounts (circles). Under the assumption that the participant obeys the axioms of expected utility theory, one can normalize the utilities of € 0 and € 80 as 0 and 1 and the vertical axis represents the utilities and the thick black line denotes the utility function of the participant. But since the thick black line contains a vertical section, the participant does not obey the axioms of expected utility theory.

Figure 14: "Utilities" deduced from EUT for participant 14 in session 2.

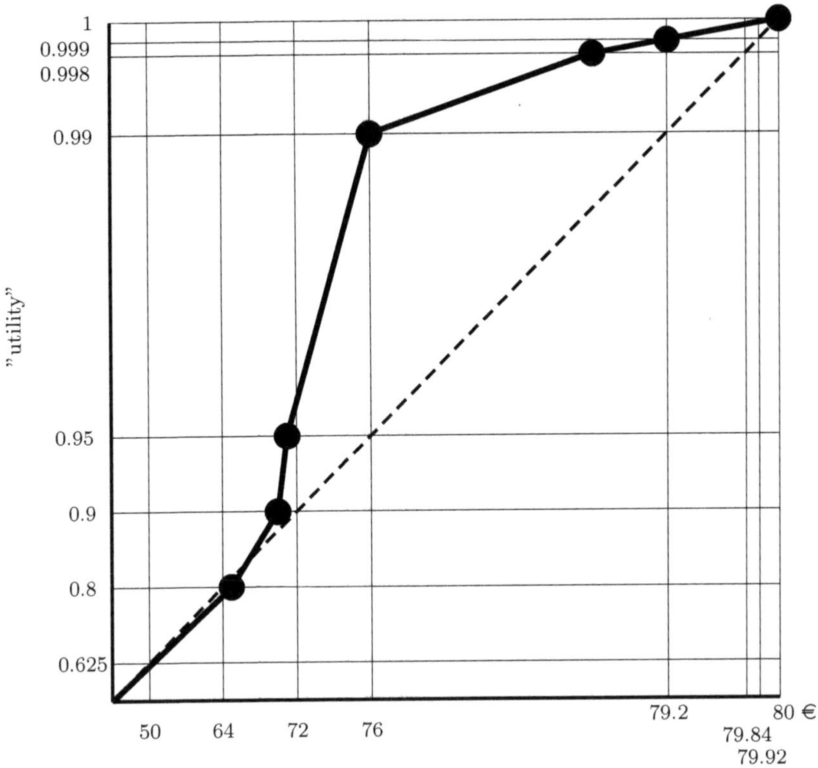

- - - - linear NM utility or fair bet price benchmark
——— participant 14 in session 2's EUT imputed "utilities" with linear interpolation between each imputed "utility".

Risk of attack	0.2	0.1	0.05	0.01	0.002	0.001
x denotes chose to insure	x	x	x	x	x	x
WTP in €	15	10	9	4	1.6	0.8
Compared to linear NM utility or fair bet	<	>	>	>	>	>
Certainty equivalent: 80 - WTP	65	70	71	76	78.4	79.2
Classification	Convex then Concave then Linear Inconsistency of WTP and insurance decisions					

1 The scales on the axes are non-linear, but chosen such that (subject to graph grid limits), the fair bet choice benchmark denoting linear utility under EUT, is linear

2 The vertical axis shows the probability of obtaining €80 without insurance. The thick black line shows the connection of this probablity with a certainty equivalent amount, namely €80 minus the willingness to pay for insurance at this probability, with linear interpolations between adjacent amounts (circles). Under the assumption that the participant obeys the axioms of expected utility theory, one can normalize the utilities of €0 and €80 as 0 and 1 and the vertical axis represents the utilities and the thick black line denotes the utility function of the participant.

Figure 15: "Utilities" deduced from EUT for participant 15 in session 2.

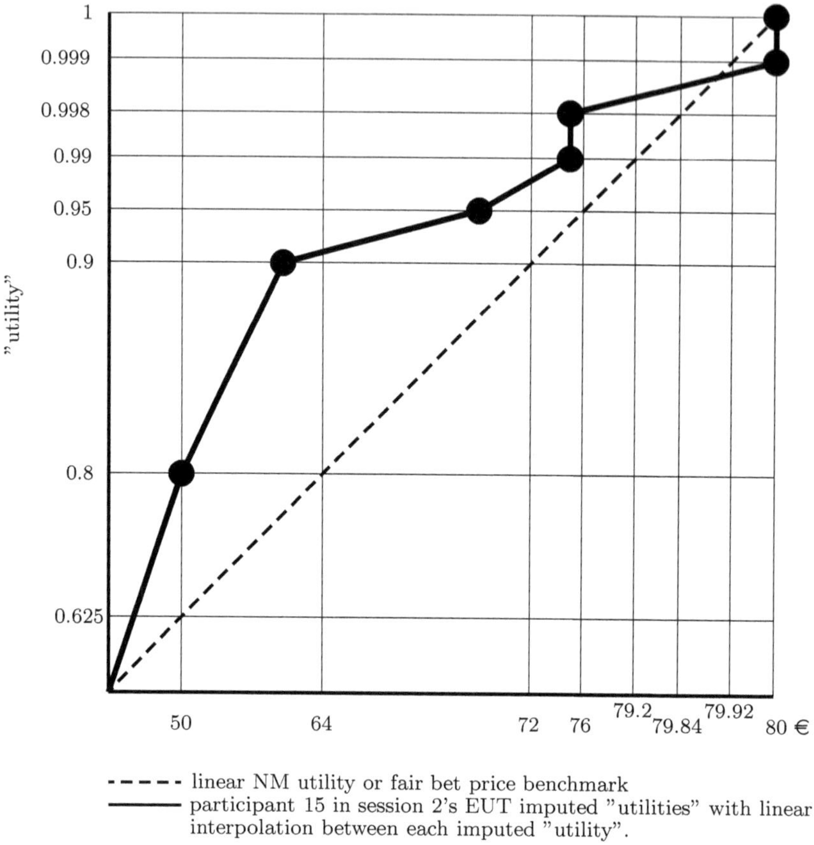

Risk of attack	0.2	0.1	0.05	0.01	0.002	0.001
x denotes chose to insure	x	x	x			
WTP in €	30	20	10	5	5	0
Compared to linear NM utility or fair bet	>	>	>	>	>	<
Certainty equivalent: 80 - WTP	50	60	70	75	75	80
Classification	Concave then Convex-Vertical then Concave then Convex-Vertical Inconsistency of WTP and insurance decisions					

1 The scales on the axes are non-linear, but chosen such that (subject to graph grid limits), the fair bet choice benchmark denoting linear utility under EUT, is linear

2 The vertical axis shows the probability of obtaining € 80 without insurance. The thick black line shows the connection of this probablity with a certainty equivalent amount, namely € 80 minus the willingness to pay for insurance at this probability, with linear interpolations between adjacent amounts (circles). Under the assumption that the participant obeys the axioms of expected utility theory, one can normalize the utilities of € 0 and € 80 as 0 and 1 and the vertical axis represents the utilities and the thick black line denotes the utility function of the participant. But since the thick black line contains vertical sections, the participant does not obey the axioms of expected utility theory.

Figure 16: "Utilities" deduced from EUT for participant 16 in session 2.

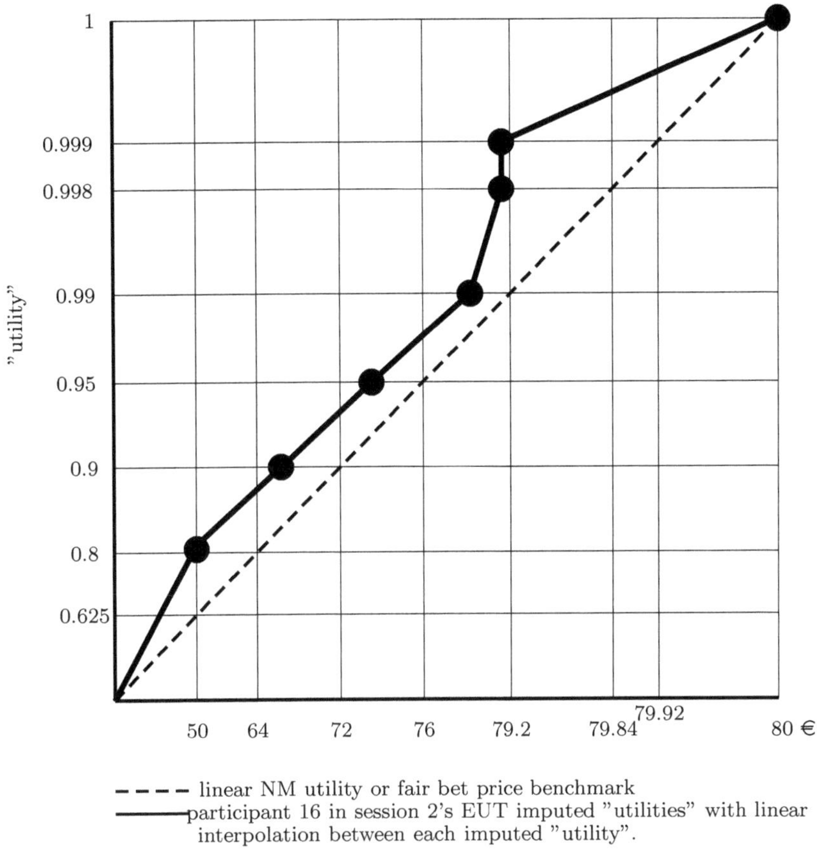

Risk of attack	0.2	0.1	0.05	0.01	0.002	0.001
x denotes chose to insure	x	x	x	x	x	x
WTP in €	30	15	7.5	1.5	1	1
Compared to linear NM utility or fair bet	>	>	>	>	>	>
Certainty equivalent: 80 - WTP	50	65	72.5	78.5	79	79
Classification	Concave then Linear then Convex-Vertical then Concave Consistency of WTP and insurance decisions					

1 The scales on the axes are non-linear, but chosen such that (subject to graph grid limits), the fair bet choice benchmark denoting linear utility under EUT, is linear

2 The vertical axis shows the probability of obtaining €80 without insurance. The thick black line shows the connection of this probablity with a certainty equivalent amount, namely €80 minus the willingness to pay for insurance at this probability, with linear interpolations between adjacent amounts (circles). Under the assumption that the participant obeys the axioms of expected utility theory, one can normalize the utilities of €0 and €80 as 0 and 1 and the vertical axis represents the utilities and the thick black line denotes the utility function of the participant. But since the thick black line contains a vertical section, the participant does not obey the axioms of expected utility theory.

Figure 17: "Utilities" deduced from EUT for participant 17 in session 2.

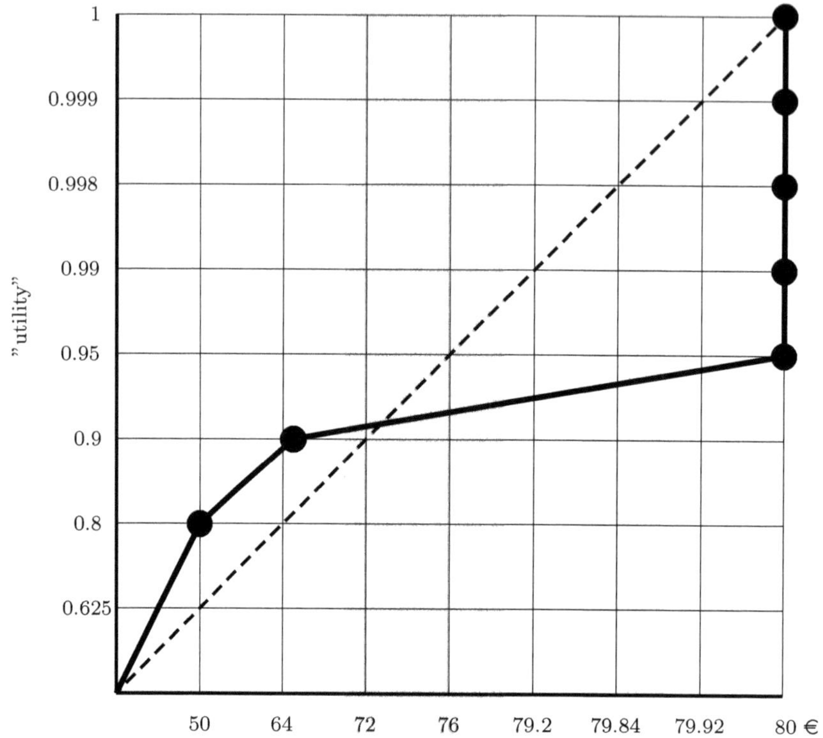

- - - - - linear NM utility or fair bet price benchmark
——— participant 17 in session 2's EUT imputed "utilities" with linear interpolation between each imputed "utility".

Risk of attack	**0.2**	**0.1**	**0.05**	**0.01**	**0.002**	**0.001**
x denotes chose to insure	x	x				
WTP in €	30	20	0	0	0	0
Compared to linear NM utility or fair bet	>	>	<	<	<	<
Certainty equivalent: 80 - WTP	50	65	80	80	80	80
Classification	Concave then Convex-Vertical Inconsistency of WTP and insurance decisions					

1 The scales on the axes are non-linear, but chosen such that (subject to graph grid limits), the fair bet choice benchmark denoting linear utility under EUT, is linear
2 The vertical axis shows the probability of obtaining €80 without insurance. The thick black line shows the connection of this probablity with a certainty equivalent amount, namely €80 minus the willingness to pay for insurance at this probability, with linear interpolations between adjacent amounts (circles). Under the assumption that the participant obeys the axioms of expected utility theory, one can normalize the utilities of €0 and €80 as 0 and 1 and the vertical axis represents the utilities and the thick black line denotes the utility function of the participant. But since the thick black line contains a vertical section, the participant does not obey the axioms of expected utility theory.

Figure 18: "Utilities" deduced from EUT for participant 18 in session 2.

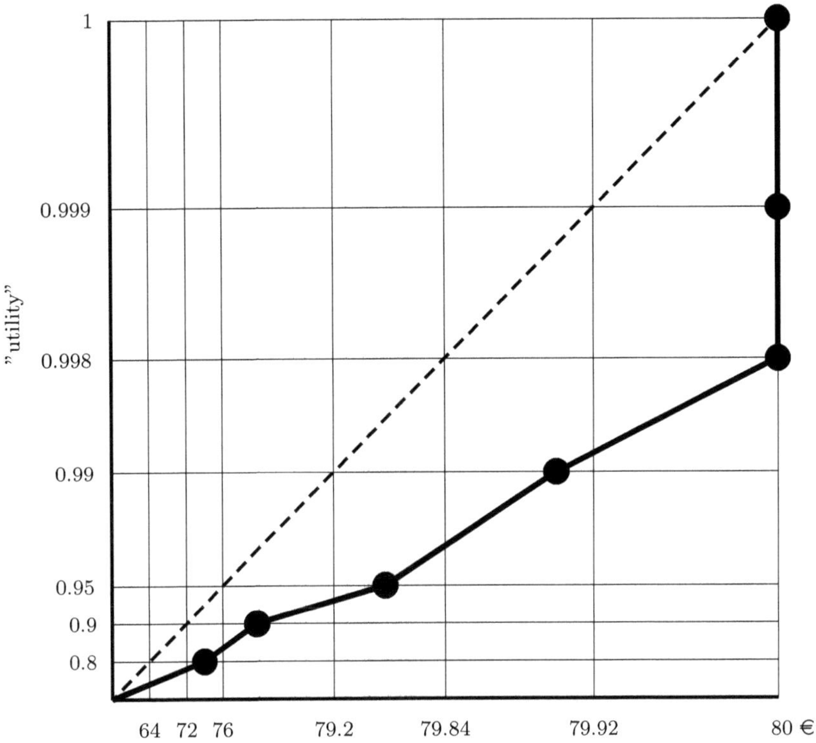

- - - - - linear NM utility or fair bet price benchmark
——— participant 18 in session 2's EUT imputed "utilities" with linear interpolation between each imputed "utility".

Risk of attack	**0.2**	**0.1**	**0.05**	**0.01**	**0.002**	**0.001**
x denotes chose to insure	x	x	x			
WTP in €	6	3	0.5	0.1	0	0
Compared to linear NM utility or fair bet	<	<	<	<	<	<
Certainty equivalent: 80 - WTP	74	77	79.5	79.9	80	80
Classification	Convex then Concave then Convex then Concave then Convex-Vertical Inconsistency of WTP and insurance decisions					

1 The scales on the axes are non-linear, but chosen such that (subject to graph grid limits), the fair bet choice benchmark denoting linear utility under EUT, is linear

2 The vertical axis shows the probability of obtaining €80 without insurance. The thick black line shows the connection of this probablity with a certainty equivalent amount, namely €80 minus the willingness to pay for insurance at this probability, with linear interpolations between adjacent amounts (circles). Under the assumption that the participant obeys the axioms of expected utility theory, one can normalize the utilities of €0 and €80 as 0 and 1 and the vertical axis represents the utilities and the thick black line denotes the utility function of the participant. But since the thick black line contains a vertical section, the participant does not obey the axioms of expected utility theory.

Figure 19: "Utilities" deduced from EUT for participant 1 in session 4.

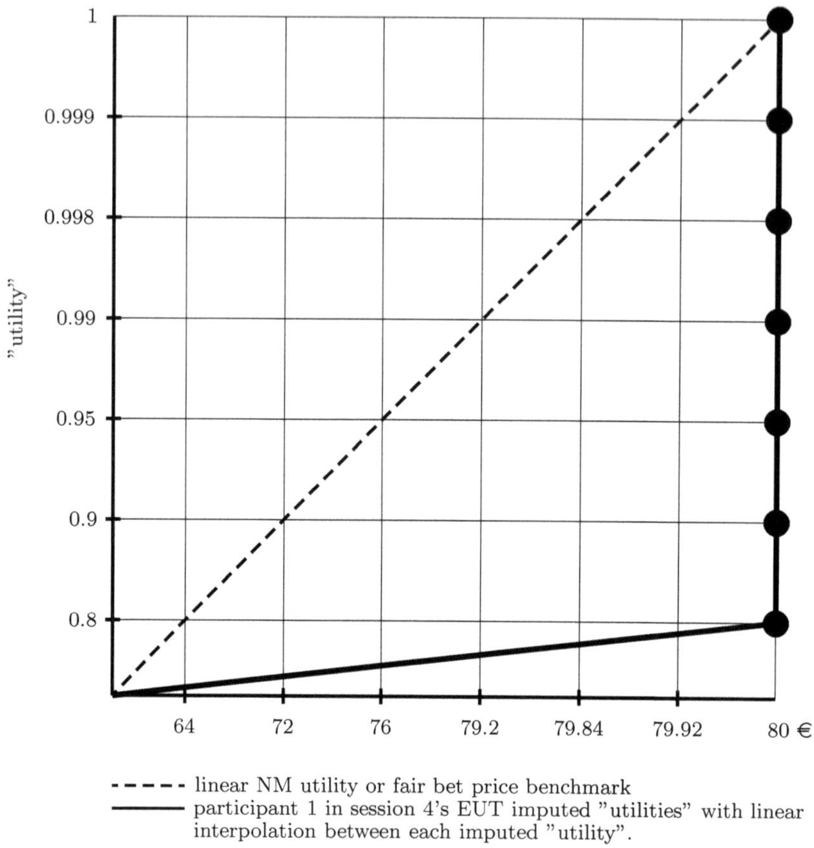

Risk of attack	0.2	0.1	0.05	0.01	0.002	0.001
x denotes chose to insure						
WTP in €	0	0	0	0	0	0
Compared to linear NM utility or fair bet	<	<	<	<	<	<
Certainty equivalent: 80 - WTP	80	80	80	80	80	80
Classification	Convex-Vertical Consistency of WTP and insurance decisions					

1 The scales on the axes are non-linear, but chosen such that (subject to graph grid limits), the fair bet choice benchmark denoting linear utility under EUT, is linear

2 The vertical axis shows the probability of obtaining €80 without insurance. The thick black line shows the connection of this probablity with a certainty equivalent amount, namely €80 minus the willingness to pay for insurance at this probability, with linear interpolations between adjacent amounts (circles). Under the assumption that the participant obeys the axioms of expected utility theory, one can normalize the utilities of €0 and €80 as 0 and 1 and the vertical axis represents the utilities and the thick black line denotes the utility function of the participant. But since the thick black line contains a vertical section, the participant does not obey the axioms of expected utility theory.

Figure 20: "Utilities" deduced from EUT for participant 3 in session 4.

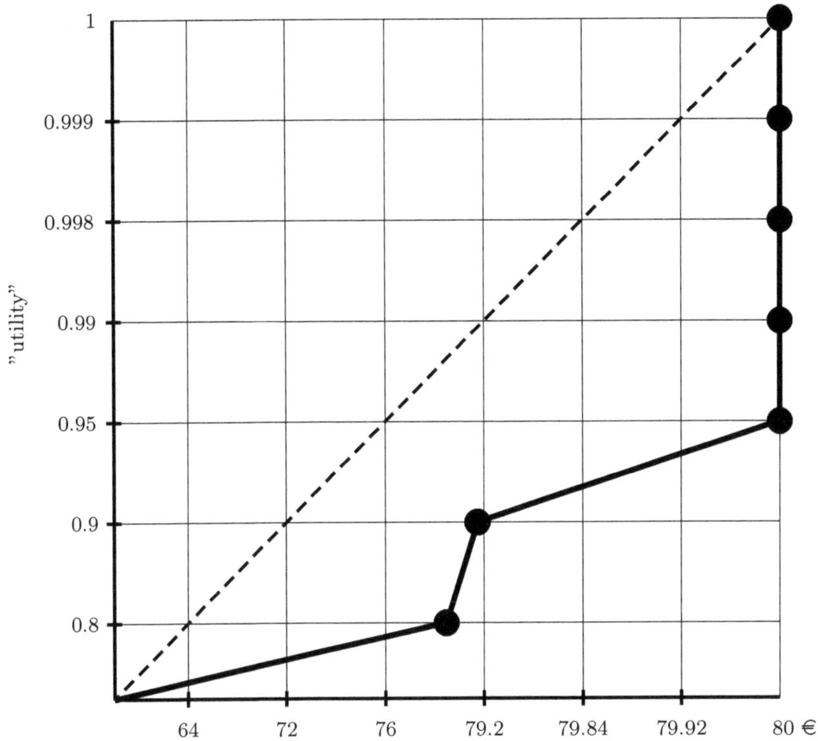

- - - - - linear NM utility or fair bet price benchmark
——— participant 3 in session 4's EUT imputed "utilities" with linear interpolation between each imputed "utility".

Risk of attack	0.2	0.1	0.05	0.01	0.002	0.001
x denotes chose to insure						
WTP in €	2	1	0	0	0	0
Compared to linear NM utility or fair bet	<	<	<	<	<	<
Certainty equivalent: 80 - WTP	78	79	80	80	80	80
Classification	Convex then Concave then Convex-Vertical Consistency of WTP and insurance decisions					

1 The scales on the axes are non-linear, but chosen such that (subject to graph grid limits), the fair bet choice benchmark denoting linear utility under EUT, is linear

2 The vertical axis shows the probability of obtaining €80 without insurance. The thick black line shows the connection of this probablity with a certainty equivalent amount, namely €80 minus the willingness to pay for insurance at this probability, with linear interpolations between adjacent amounts (circles). Under the assumption that the participant obeys the axioms of expected utility theory, one can normalize the utilities of €0 and €80 as 0 and 1 and the vertical axis represents the utilities and the thick black line denotes the utility function of the participant. But since the thick black line contains a vertical section, the participant does not obey the axioms of expected utility theory.

Figure 21: "Utilities" deduced from EUT for participant 4 in session 4.

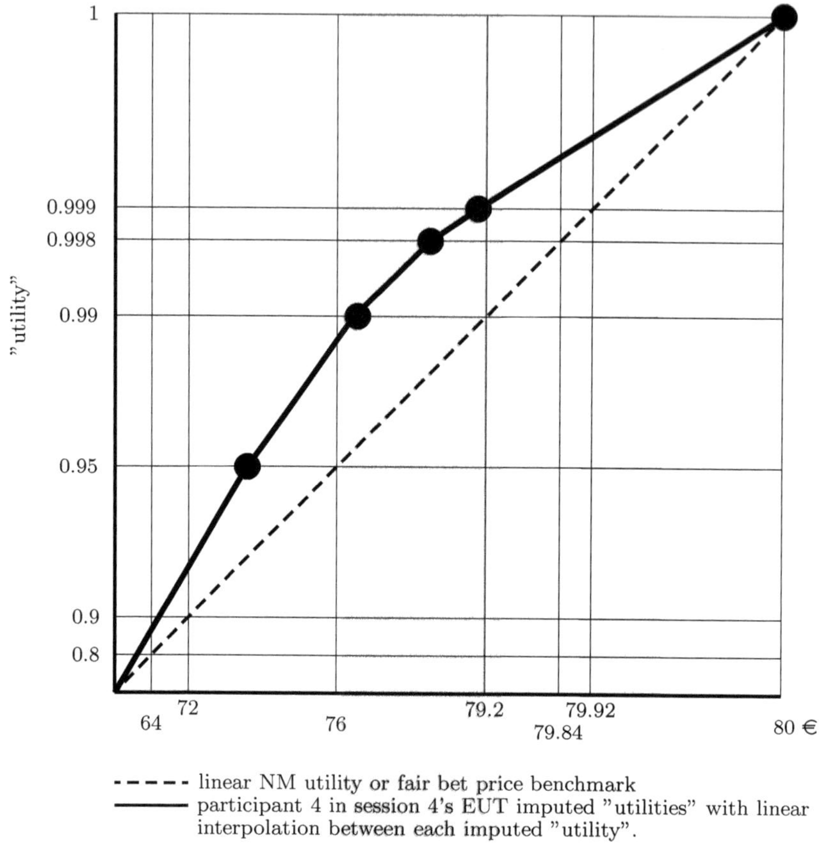

Risk of attack	0.2	0.1	0.05	0.01	0.002	0.001
x denotes chose to insure	x	x				
WTP in €	E	E	6	3	2	1
Compared to linear NM utility or fair bet	E	E	>	>	>	>
Certainty equivalent: 80 - WTP	E	E	74	77	78	79
Classification	Concave then Linear Consistency of WTP and insurance decisions					

E denotes that the participants left these fields empty in the questionnaire

1 The scales on the axes are non-linear, but chosen such that (subject to graph grid limits), the fair bet choice benchmark denoting linear utility under EUT, is linear

2 The vertical axis shows the probability of obtaining € 80 without insurance. The thick black line shows the connection of this probablity with a certainty equivalent amount, namely € 80 minus the willingness to pay for insurance at this probability, with linear interpolations between adjacent amounts (circles). Under the assumption that the participant obeys the axioms of expected utility theory, one can normalize the utilities of € 0 and € 80 as 0 and 1 and the vertical axis represents the utilities and the thick black line denotes the utility function of the participant.

Figure 22: "Utilities" deduced from EUT for participant 5 in session 4.

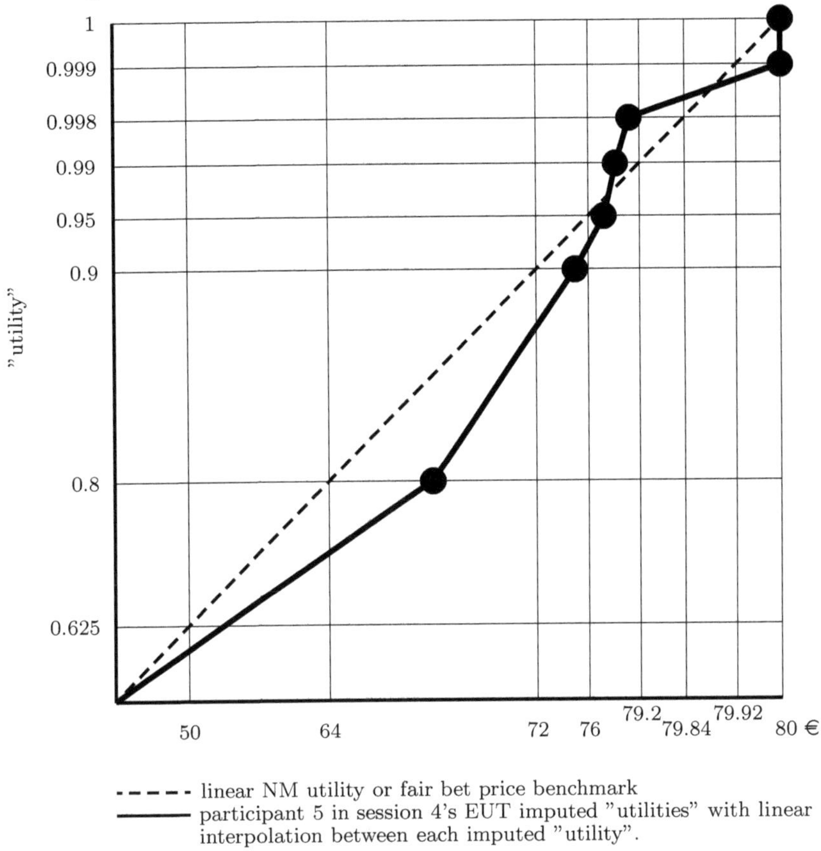

Risk of attack	0.2	0.1	0.05	0.01	0.002	0.001
x denotes chose to insure	x					
WTP in €	12	5	3	2	1	0
Compared to linear NM utility or fair bet	<	<	<	>	>	<
Certainty equivalent: 80 - WTP	68	75	77	78	79	80
Classification	Convex then Concave then Convex-Vertical Consistency of WTP and insurance decisions					

1 The scales on the axes are non-linear, but chosen such that (subject to graph grid limits), the fair bet choice benchmark denoting linear utility under EUT, is linear

2 The vertical axis shows the probability of obtaining €80 without insurance. The thick black line shows the connection of this probablity with a certainty equivalent amount, namely €80 minus the willingness to pay for insurance at this probability, with linear interpolations between adjacent amounts (circles). Under the assumption that the participant obeys the axioms of expected utility theory, one can normalize the utilities of €0 and €80 as 0 and 1 and the vertical axis represents the utilities and the thick black line denotes the utility function of the participant. But since the thick black line contains a vertical section, the participant does not obey the axioms of expected utility theory.

Figure 23: "Utilities" deduced from EUT for participant 6 in session 4.

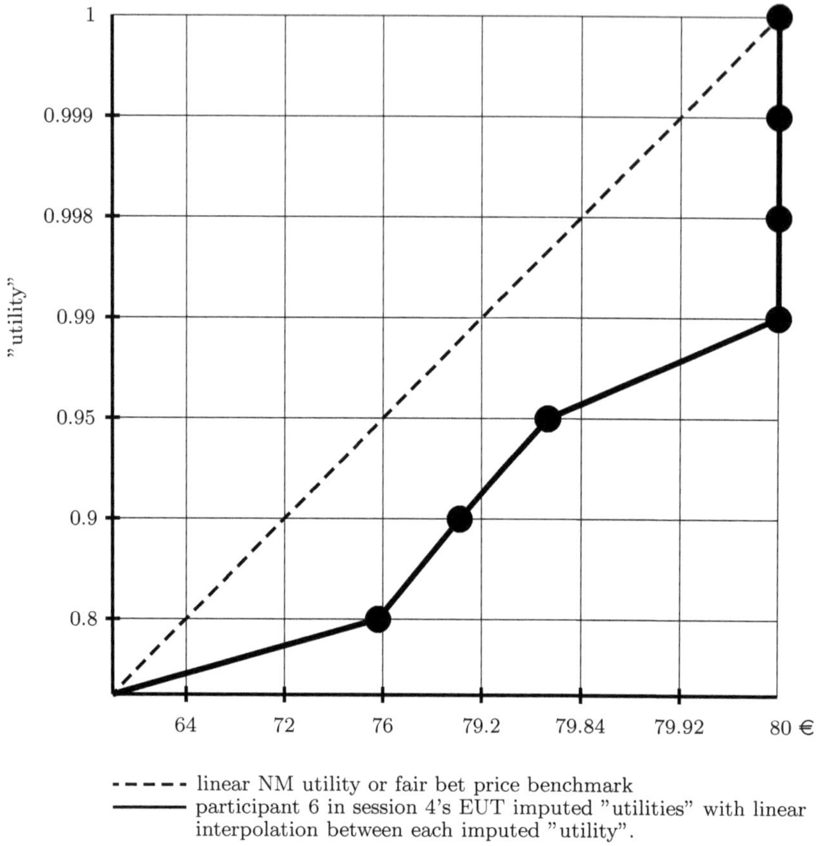

Risk of attack	0.2	0.1	0.05	0.01	0.002	0.001
x denotes chose to insure						
WTP in €	4	2	1	0	0	0
Compared to linear NM utility or fair bet	<	<	<	<	<	<
Certainty equivalent: 80 - WTP	76	78	79	80	80	80
Classification	Convex then Linear then Concave then Convex-Vertical Consistency of WTP and insurance decisions					

1 The scales on the axes are non-linear, but chosen such that (subject to graph grid limits), the fair bet choice benchmark denoting linear utility under EUT, is linear

2 The vertical axis shows the probability of obtaining €80 without insurance. The thick black line shows the connection of this probablity with a certainty equivalent amount, namely €80 minus the willingness to pay for insurance at this probability, with linear interpolations between adjacent amounts (circles). Under the assumption that the participant obeys the axioms of expected utility theory, one can normalize the utilities of €0 and €80 as 0 and 1 and the vertical axis represents the utilities and the thick black line denotes the utility function of the participant. But since the thick black line contains a vertical section, the participant does not obey the axioms of expected utility theory.

Figure 24: "Utilities" deduced from EUT for participant 7 in session 4.

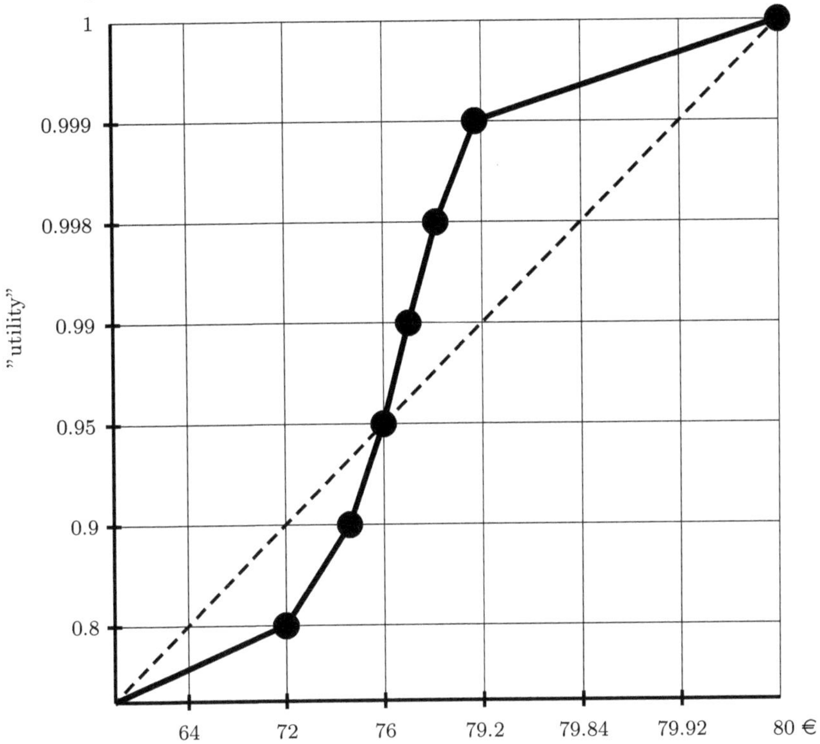

- - - - - linear NM utility or fair bet price benchmark
——— participant 7 in session 4's EUT imputed "utilities" with linear interpolation between each imputed "utility".

Risk of attack	0.2	0.1	0.05	0.01	0.002	0.001
x denotes chose to insure						
WTP in €	8	6	4	2	1.5	1
Compared to linear NM utility or fair bet	<	<	=	>	>	>
Certainty equivalent: 80 - WTP	72	74	76	78	78.5	79
Classification	Convex then Concave Consistency of WTP and insurance decisions					

1 The scales on the axes are non-linear, but chosen such that (subject to graph grid limits), the fair bet choice benchmark denoting linear utility under EUT, is linear

2 The vertical axis shows the probability of obtaining € 80 without insurance. The thick black line shows the connection of this probablity with a certainty equivalent amount, namely € 80 minus the willingness to pay for insurance at this probability, with linear interpolations between adjacent amounts (circles). Under the assumption that the participant obeys the axioms of expected utility theory, one can normalize the utilities of € 0 and € 80 as 0 and 1 and the vertical axis represents the utilities and the thick black line denotes the utility function of the participant.

Figure 25: "Utilities" deduced from EUT for participant 8 in session 4.

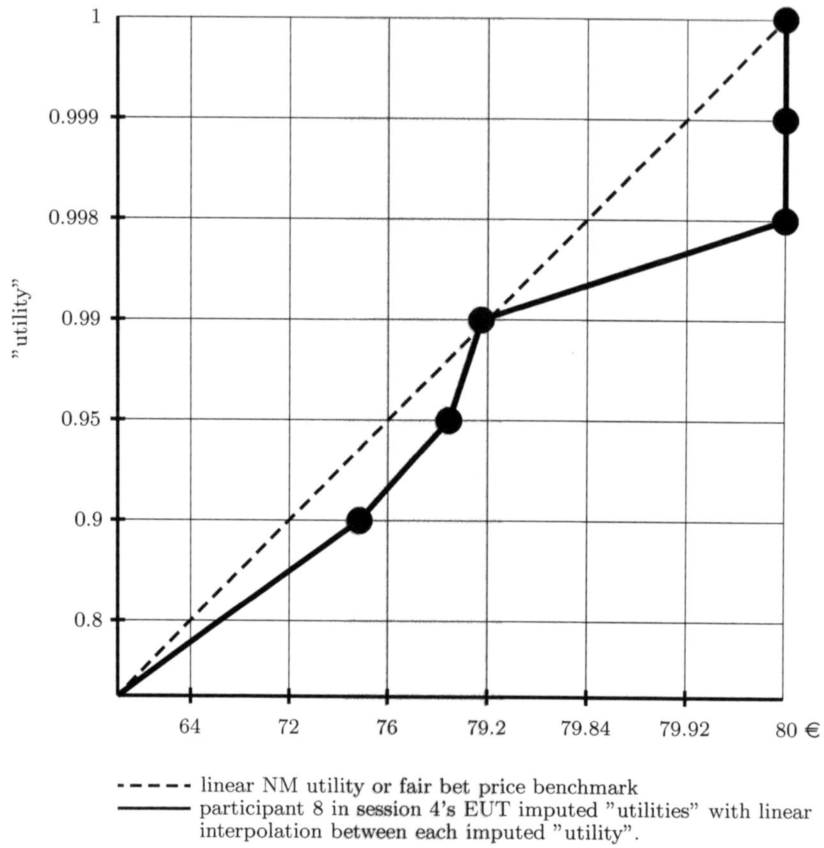

Risk of attack	0.2	0.1	0.05	0.01	0.002	0.001
x denotes chose to insure	x					
WTP in €	E	5	2	1	0	0
Compared to linear NM utility or fair bet	E	<	<	>	<	<
Certainty equivalent: 80 - WTP	E	75	78	79	80	80
Classification	Convex then Concave then Convex-Vertical Consistency of WTP and insurance decisions					

E denotes that the participants left these fields empty in the questionnaire

1 The scales on the axes are non-linear, but chosen such that (subject to graph grid limits), the fair bet choice benchmark denoting linear utility under EUT, is linear
2 The vertical axis shows the probability of obtaining € 80 without insurance. The thick black line shows the connection of this probablity with a certainty equivalent amount, namely € 80 minus the willingness to pay for insurance at this probability, with linear interpolations between adjacent amounts (circles). Under the assumption that the participant obeys the axioms of expected utility theory, one can normalize the utilities of € 0 and € 80 as 0 and 1 and the vertical axis represents the utilities and the thick black line denotes the utility function of the participant. But since the thick black line contains a vertical section, the participant does not obey the axioms of expected utility theory.

Figure 26: "Utilities" deduced from EUT for participant 10 in session 4.

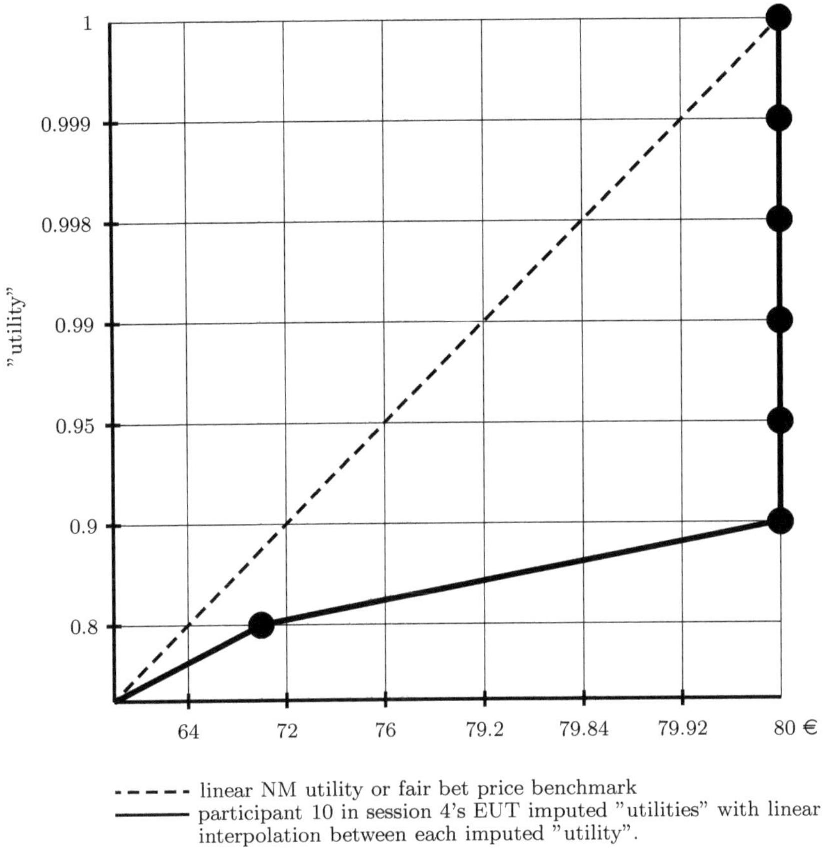

Risk of attack	**0.2**	**0.1**	**0.05**	**0.01**	**0.002**	**0.001**
x denotes chose to insure						
WTP in €	10	0	0	0	0	0
Compared to linear NM utility or fair bet	<	<	<	<	<	<
Certainty equivalent: 80 - WTP	70	80	80	80	80	80
Classification	Concave then Convex-Vertical Consistency of WTP and insurance decisions					

1 The scales on the axes are non-linear, but chosen such that (subject to graph grid limits), the fair bet choice benchmark denoting linear utility under EUT, is linear

2 The vertical axis shows the probability of obtaining €80 without insurance. The thick black line shows the connection of this probablity with a certainty equivalent amount, namely €80 minus the willingness to pay for insurance at this probability, with linear interpolations between adjacent amounts (circles). Under the assumption that the participant obeys the axioms of expected utility theory, one can normalize the utilities of €0 and €80 as 0 and 1 and the vertical axis represents the utilities and the thick black line denotes the utility function of the participant. But since the thick black line contains a vertical section, the participant does not obey the axioms of expected utility theory.

Figure 27: "Utilities" deduced from EUT for participant 11 in session 4.

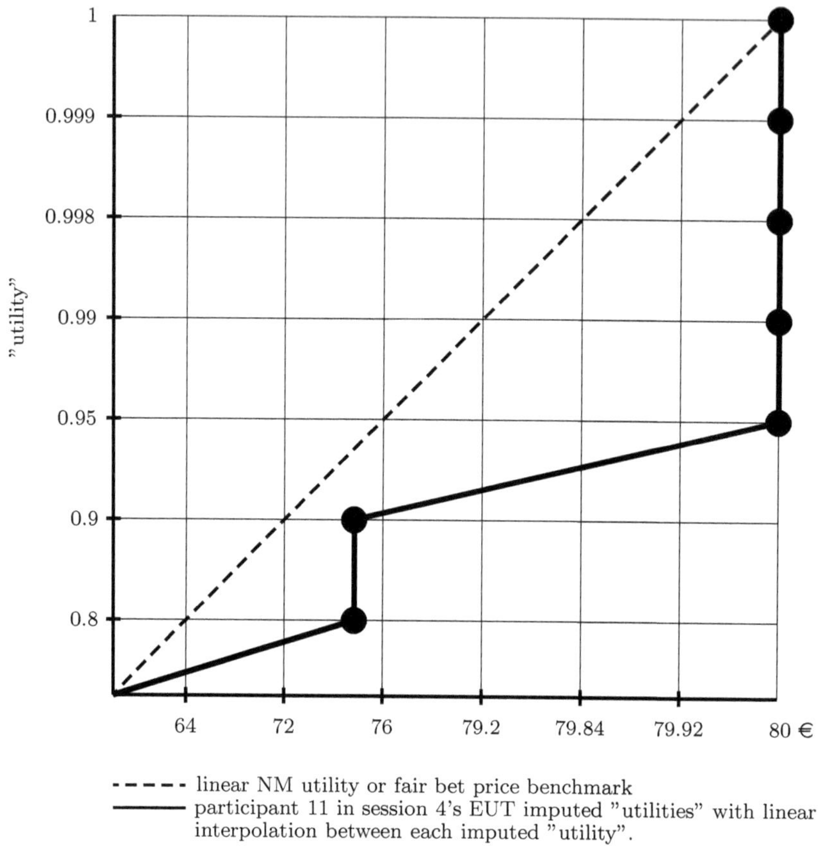

Risk of attack	0.2	0.1	0.05	0.01	0.002	0.001
x denotes chose to insure						
WTP in €	5	5	0	0	0	0
Compared to linear NM utility or fair bet	<	<	<	<	<	<
Certainty equivalent: 80 - WTP	75	75	80	80	80	80
Classification	Convex-Vertical then Concave then Convex-Vertical Consistency of WTP and insurance decisions					

1 The scales on the axes are non-linear, but chosen such that (subject to graph grid limits), the fair bet choice benchmark denoting linear utility under EUT, is linear

2 The vertical axis shows the probability of obtaining € 80 without insurance. The thick black line shows the connection of this probablity with a certainty equivalent amount, namely € 80 minus the willingness to pay for insurance at this probability, with linear interpolations between adjacent amounts (circles). Under the assumption that the participant obeys the axioms of expected utility theory, one can normalize the utilities of € 0 and € 80 as 0 and 1 and the vertical axis represents the utilities and the thick black line denotes the utility function of the participant. But since the thick black line contains vertical sections, the participant does not obey the axioms of expected utility theory.

Figure 28: "Utilities" deduced from EUT for participant 13 in session 4.

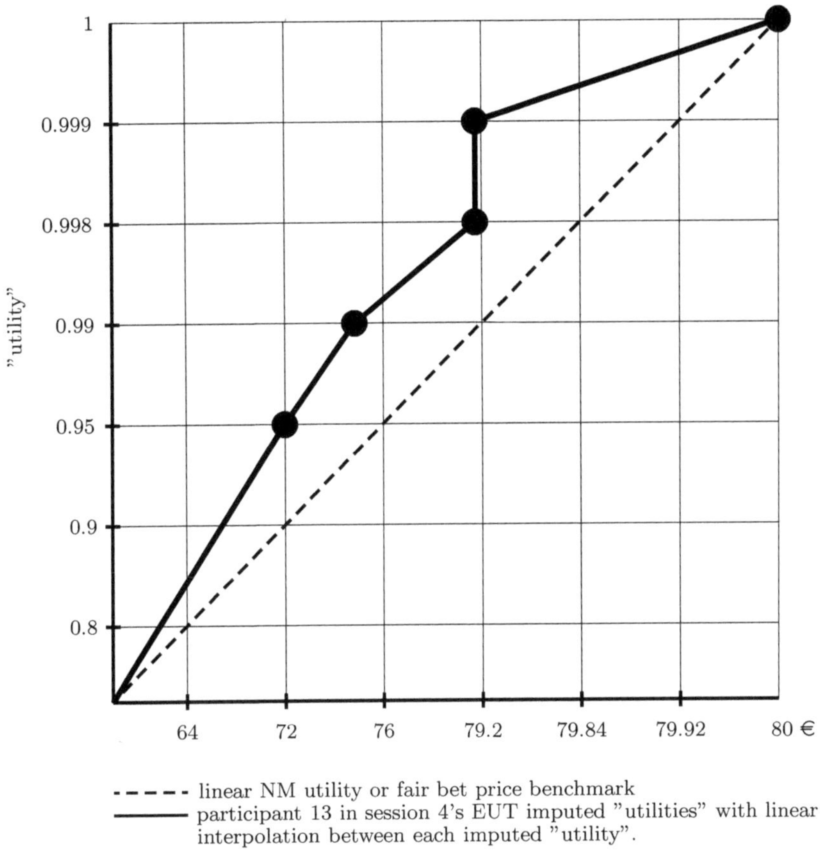

Risk of attack	0.2	0.1	0.05	0.01	0.002	0.001
x denotes chose to insure	x	x				
WTP in €	E	E	8	5	1	1
Compared to linear NM utility or fair bet	E	E	>	>	>	>
Certainty equivalent: 80 - WTP	E	E	72	75	79	79
Classification	Concave then Convex-Vertical then Concave Consistency of WTP and insurance decisions					

E denotes that the participants left these fields empty in the questionnaire

1 The scales on the axes are non-linear, but chosen such that (subject to graph grid limits), the fair bet choice benchmark denoting linear utility under EUT, is linear

2 The vertical axis shows the probability of obtaining €80 without insurance. The thick black line shows the connection of this probablity with a certainty equivalent amount, namely €80 minus the willingness to pay for insurance at this probability, with linear interpolations between adjacent amounts (circles). Under the assumption that the participant obeys the axioms of expected utility theory, one can normalize the utilities of €0 and €80 as 0 and 1 and the vertical axis represents the utilities and the thick black line denotes the utility function of the participant. But since the thick black line contains a vertical section, the participant does not obey the axioms of expected utility theory.

Figure 29: "Utilities" deduced from EUT for participant 14 in session 4.

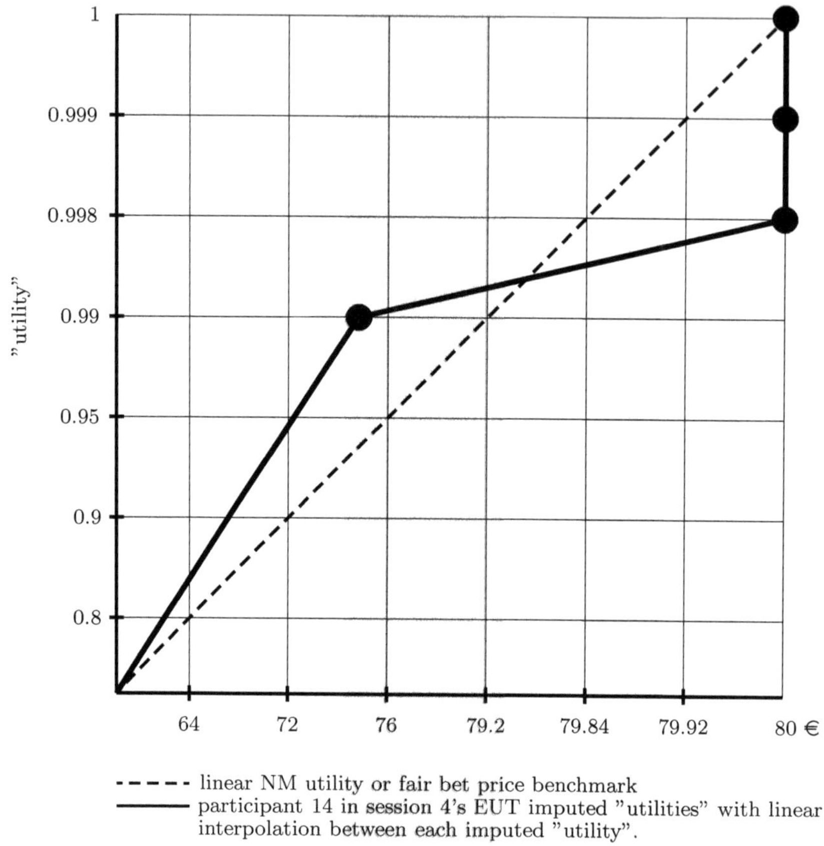

Risk of attack	**0.2**	**0.1**	**0.05**	**0.01**	**0.002**	**0.001**
x denotes chose to insure	x	x	x			
WTP in €	E	E	E	5	0	0
Compared to linear NM utility or fair bet	E	E	E	>	<	<
Certainty equivalent: 80 - WTP	E	E	E	75	80	80
Classification	Concave then Convex-Vertical Consistency of WTP and insurance decisions					

E denotes that the participants left these fields empty in the questionnaire

1 The scales on the axes are non-linear, but chosen such that (subject to graph grid limits), the fair bet choice benchmark denoting linear utility under EUT, is linear
2 The vertical axis shows the probability of obtaining €80 without insurance. The thick black line shows the connection of this probablity with a certainty equivalent amount, namely €80 minus the willingness to pay for insurance at this probability, with linear interpolations between adjacent amounts (circles). Under the assumption that the participant obeys the axioms of expected utility theory, one can normalize the utilities of €0 and €80 as 0 and 1 and the vertical axis represents the utilities and the thick black line denotes the utility function of the participant. But since the thick black line contains a vertical section, the participant does not obey the axioms of expected utility theory.

Figure 30: "Utilities" deduced from EUT for participant 15 in session 4.

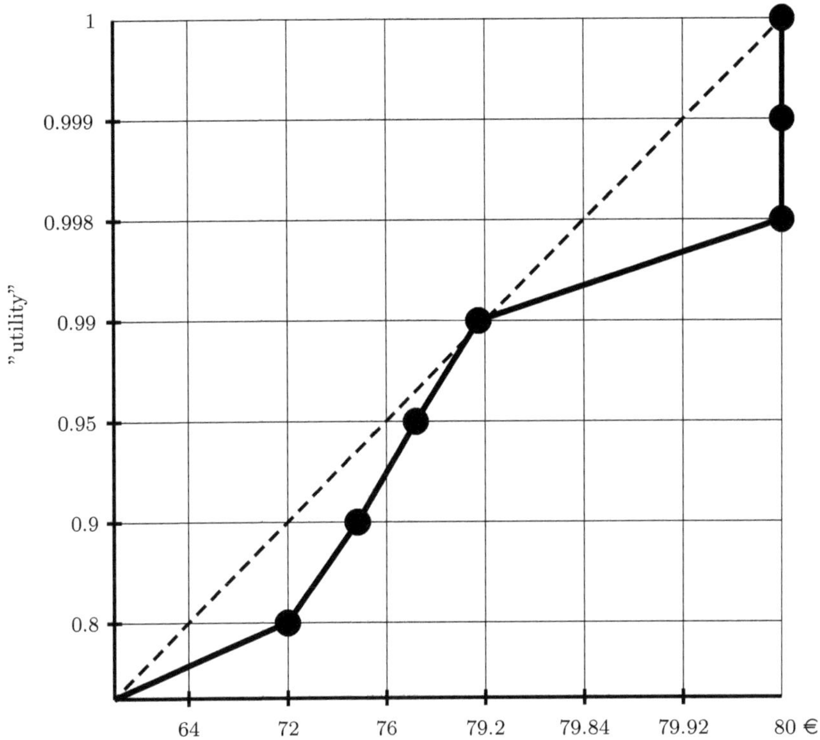

- - - - - linear NM utility or fair bet price benchmark
——— participant 15 in session 4's EUT imputed "utilities" with linear interpolation between each imputed "utility".

Risk of attack	**0.2**	**0.1**	**0.05**	**0.01**	**0.002**	**0.001**
x denotes chose to insure						
WTP in €	8	5	3	1	0	0
Compared to linear NM utility or fair bet	<	<	<	>	<	<
Certainty equivalent: 80 - WTP	72	75	77	79	80	80
Classification	Convex then Concave then Convex-Vertical Consistency of WTP and insurance decisions					

1 The scales on the axes are non-linear, but chosen such that (subject to graph grid limits), the fair bet choice benchmark denoting linear utility under EUT, is linear
2 The vertical axis shows the probability of obtaining €80 without insurance. The thick black line shows the connection of this probablity with a certainty equivalent amount, namely €80 minus the willingness to pay for insurance at this probability, with linear interpolations between adjacent amounts (circles). Under the assumption that the participant obeys the axioms of expected utility theory, one can normalize the utilities of €0 and €80 as 0 and 1 and the vertical axis represents the utilities and the thick black line denotes the utility function of the participant. But since the thick black line contains a vertical section, the participant does not obey the axioms of expected utility theory.

Figure 31: "Utilities" deduced from EUT for participant 18 in session 4.

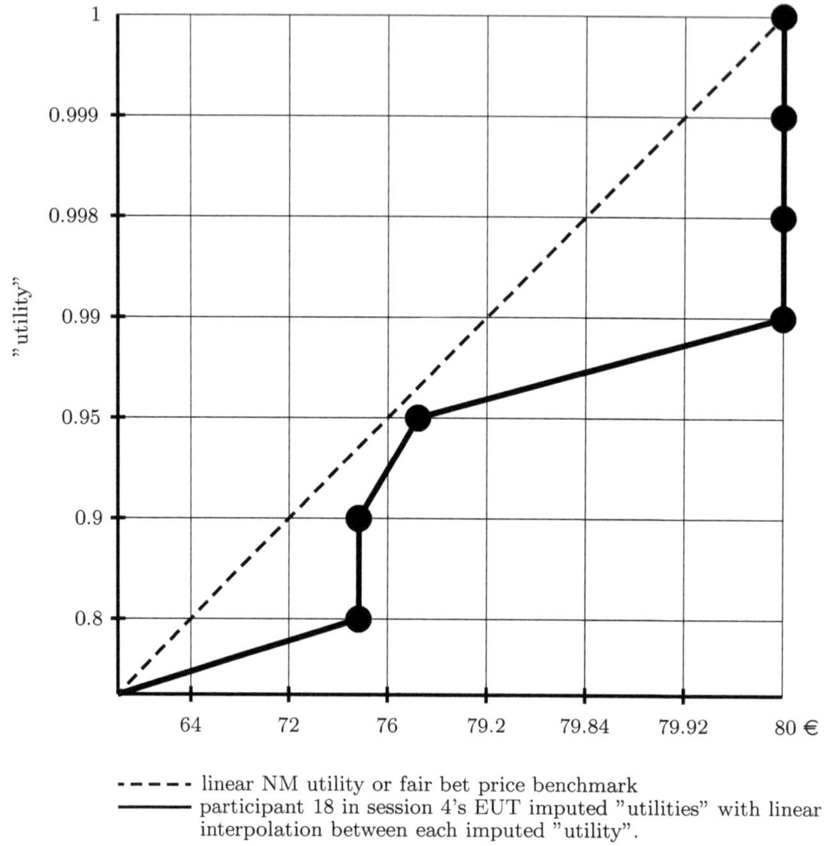

Risk of attack	0.2	0.1	0.05	0.01	0.002	0.001
x denotes chose to insure						
WTP in €	5	5	3	0	0	0
Compared to linear NM utility or fair bet	<	<	<	<	<	<
Certainty equivalent: 80 - WTP	75	75	77	80	80	80
Classification	Convex-Vertical then Concave then Convex-Vertical Consistency of WTP and insurance decisions					

1 The scales on the axes are non-linear, but chosen such that (subject to graph grid limits), the fair bet choice benchmark denoting linear utility under EUT, is linear

2 The vertical axis shows the probability of obtaining €80 without insurance. The thick black line shows the connection of this probablity with a certainty equivalent amount, namely €80 minus the willingness to pay for insurance at this probability, with linear interpolations between adjacent amounts (circles). Under the assumption that the participant obeys the axioms of expected utility theory, one can normalize the utilities of €0 and €80 as 0 and 1 and the vertical axis represents the utilities and the thick black line denotes the utility function of the participant. But since the thick black line contains vertical sections, the participant does not obey the axioms of expected utility theory.

Figure 32: "Utilities" deduced from EUT for participant 19 in session 4.

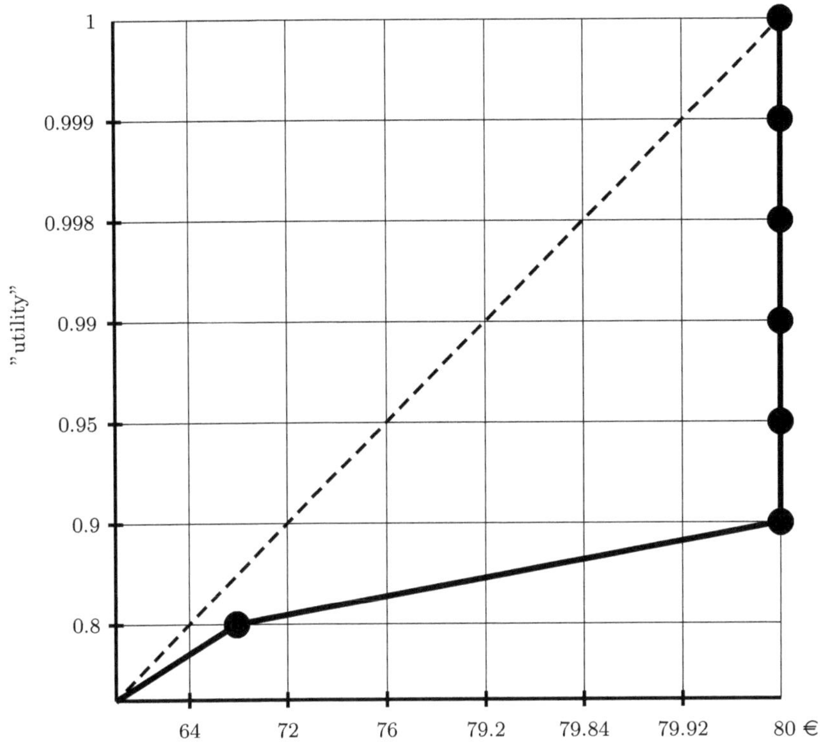

- - - - - linear NM utility or fair bet price benchmark
——— participant 19 in session 4's EUT imputed "utilities" with linear interpolation between each imputed "utility".

Risk of attack	**0.2**	**0.1**	**0.05**	**0.01**	**0.002**	**0.001**
x denotes chose to insure						
WTP in €	12	0	0	0	0	0
Compared to linear NM utility or fair bet	<	<	<	<	<	<
Certainty equivalent: 80 - WTP	68	80	80	80	80	80
Classification	Concave then Convex-Vertical Consistency of WTP and insurance decisions					

1 The scales on the axes are non-linear, but chosen such that (subject to graph grid limits), the fair bet choice benchmark denoting linear utility under EUT, is linear

2 The vertical axis shows the probability of obtaining €80 without insurance. The thick black line shows the connection of this probablity with a certainty equivalent amount, namely €80 minus the willingness to pay for insurance at this probability, with linear interpolations between adjacent amounts (circles). Under the assumption that the participant obeys the axioms of expected utility theory, one can normalize the utilities of €0 and €80 as 0 and 1 and the vertical axis represents the utilities and the thick black line denotes the utility function of the participant. But since the thick black line contains a vertical section, the participant does not obey the axioms of expected utility theory.

Figure 33: "Utilities" deduced from EUT for participant 1 in session 5.

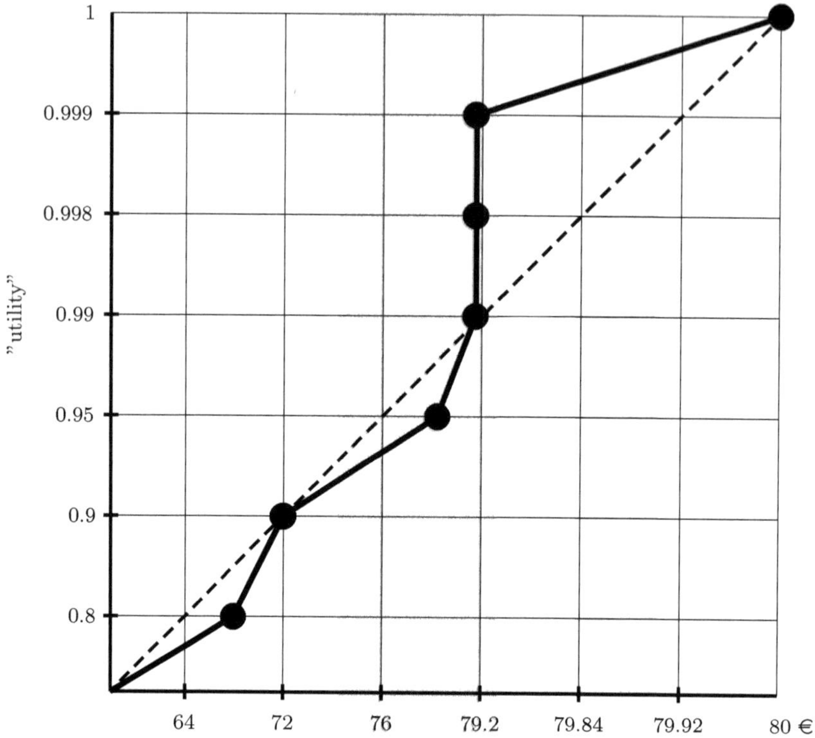

- - - - linear NM utility or fair bet price benchmark
——— participant 1 in session 5's EUT imputed "utilities" with linear interpolation between each imputed "utility".

Risk of attack	0.2	0.1	0.05	0.01	0.002	0.001
x denotes chose to insure						
WTP in €	12	8	2	1	1	1
Compared to linear NM utility or fair bet	<	=	<	>	>	>
Certainty equivalent: 80 - WTP	68	72	78	79	79	79
Classification	Convex then Concave then Convex-Vertical then Concave Inconsistency of WTP and insurance decisions					

1 The scales on the axes are non-linear, but chosen such that (subject to graph grid limits), the fair bet choice benchmark denoting linear utility under EUT, is linear

2 The vertical axis shows the probability of obtaining €80 without insurance. The thick black line shows the connection of this probablity with a certainty equivalent amount, namely €80 minus the willingness to pay for insurance at this probability, with linear interpolations between adjacent amounts (circles). Under the assumption that the participant obeys the axioms of expected utility theory, one can normalize the utilities of €0 and €80 as 0 and 1 and the vertical axis represents the utilities and the thick black line denotes the utility function of the participant. But since the thick black line contains a vertical section, the participant does not obey the axioms of expected utility theory.

Figure 34: "Utilities" deduced from EUT for participant 2 in session 5.

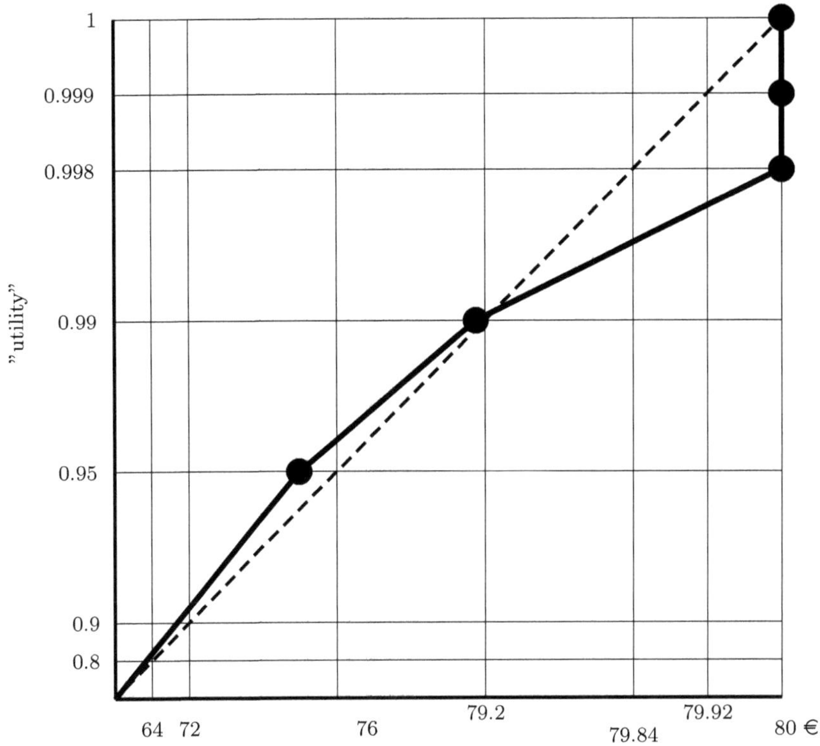

- - - - - linear NM utility or fair bet price benchmark
——— participant 2 in session 5's EUT imputed "utilities" with linear interpolation between each imputed "utility".

Risk of attack	**0.2**	**0.1**	**0.05**	**0.01**	**0.002**	**0.001**
x denotes chose to insure	x	x				
WTP in €	E	E	5	1	0	0
Compared to linear NM utility or fair bet	E	E	>	>	<	<
Certainty equivalent: 80 - WTP	E	E	75	79	80	80
Classification	Concave then Convex-Vertical Consistency of WTP and insurance decisions					

E denotes that the participants left these fields empty in the questionnaire

1 The scales on the axes are non-linear, but chosen such that (subject to graph grid limits), the fair bet choice benchmark denoting linear utility under EUT, is linear

2 The vertical axis shows the probability of obtaining €80 without insurance. The thick black line shows the connection of this probablity with a certainty equivalent amount, namely €80 minus the willingness to pay for insurance at this probability, with linear interpolations between adjacent amounts (circles). Under the assumption that the participant obeys the axioms of expected utility theory, one can normalize the utilities of €0 and €80 as 0 and 1 and the vertical axis represents the utilities and the thick black line denotes the utility function of the participant. But since the thick black line contains a vertical section, the participant does not obey the axioms of expected utility theory.

Figure 35: "Utilities" deduced from EUT for participant 3 in session 5.

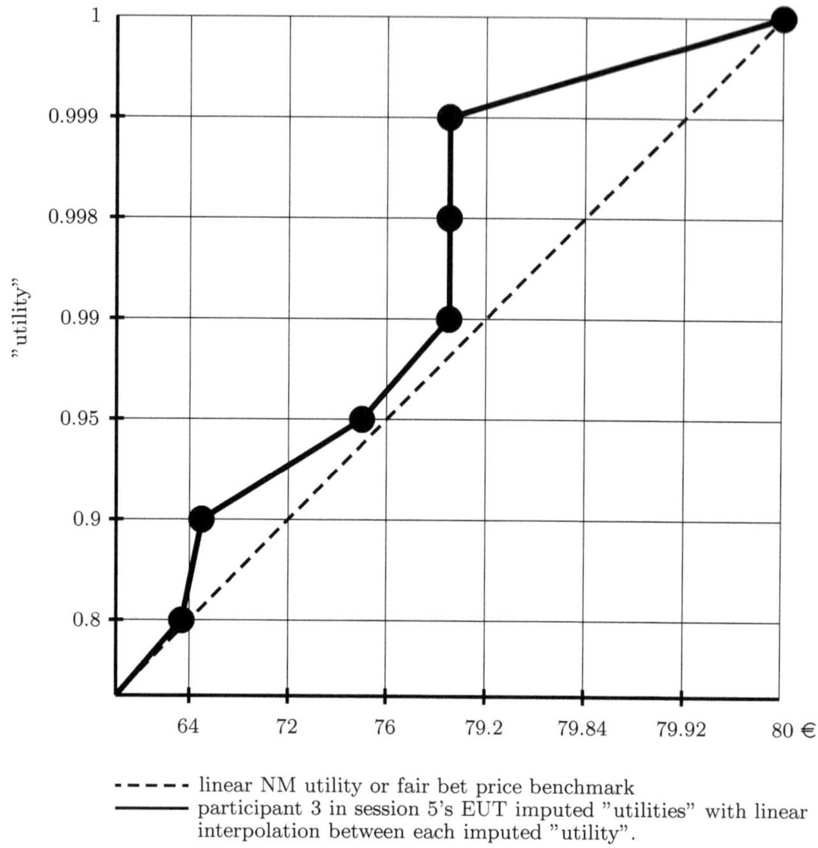

Risk of attack	0.2	0.1	0.05	0.01	0.002	0.001
x denotes chose to insure						
WTP in €	17	15	5	2	2	2
Compared to linear NM utility or fair bet	>	>	>	>	>	>
Certainty equivalent: 80 - WTP	63	65	75	78	78	78
Classification	Convex then Concave then Convex-Vertical then Concave Inconsistency of WTP and insurance decisions					

1 The scales on the axes are non-linear, but chosen such that (subject to graph grid limits), the fair bet choice benchmark denoting linear utility under EUT, is linear
2 The vertical axis shows the probability of obtaining € 80 without insurance. The thick black line shows the connection of this probablity with a certainty equivalent amount, namely € 80 minus the willingness to pay for insurance at this probability, with linear interpolations between adjacent amounts (circles). Under the assumption that the participant obeys the axioms of expected utility theory, one can normalize the utilities of € 0 and € 80 as 0 and 1 and the vertical axis represents the utilities and the thick black line denotes the utility function of the participant. But since the thick black line contains a vertical section, the participant does not obey the axioms of expected utility theory.

Figure 36: "Utilities" deduced from EUT for participant 4 in session 5.

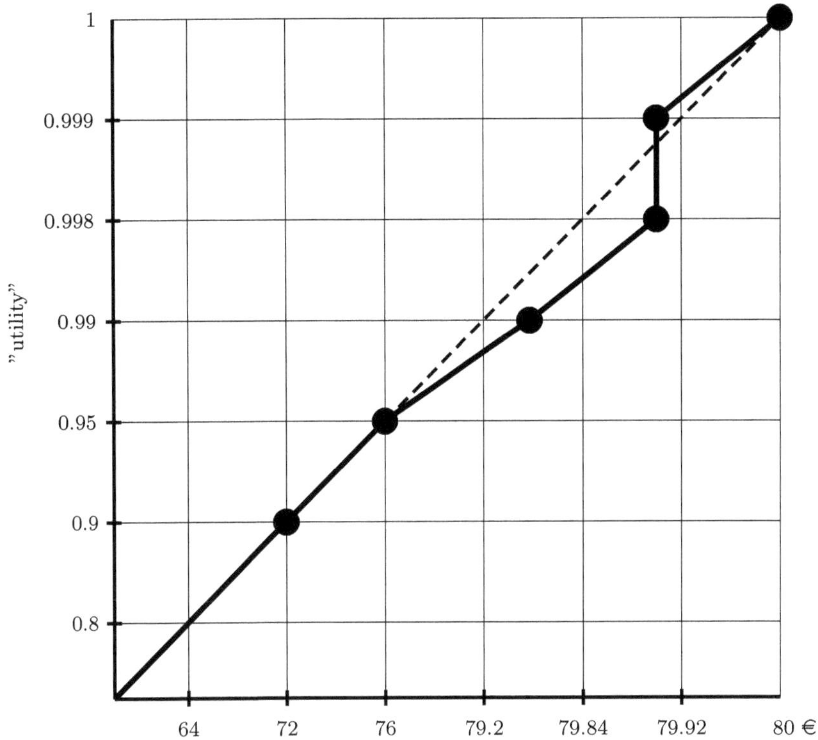

- - - - - linear NM utility or fair bet price benchmark
——— participant 4 in session 5's EUT imputed "utilities" with linear interpolation between each imputed "utility".

Risk of attack	**0.2**	**0.1**	**0.05**	**0.01**	**0.002**	**0.001**
x denotes chose to insure	x					
WTP in €	E	8	4	0.5	0.1	0.1
Compared to linear NM utility or fair bet	E	=	=	<	<	>
Certainty equivalent: 80 - WTP	E	72	76	79.5	79.9	79.9
Classification	Concave then Convex-Vertical then Concave Consistency of WTP and insurance decisions					

E denotes that the participants left these fields empty in the questionnaire

1 The scales on the axes are non-linear, but chosen such that (subject to graph grid limits), the fair bet choice benchmark denoting linear utility under EUT, is linear

2 The vertical axis shows the probability of obtaining €80 without insurance. The thick black line shows the connection of this probablity with a certainty equivalent amount, namely €80 minus the willingness to pay for insurance at this probability, with linear interpolations between adjacent amounts (circles). Under the assumption that the participant obeys the axioms of expected utility theory, one can normalize the utilities of €0 and €80 as 0 and 1 and the vertical axis represents the utilities and the thick black line denotes the utility function of the participant. But since the thick black line contains a vertical section, the participant does not obey the axioms of expected utility theory.

Figure 37: "Utilities" deduced from EUT for participant 5 in session 5.

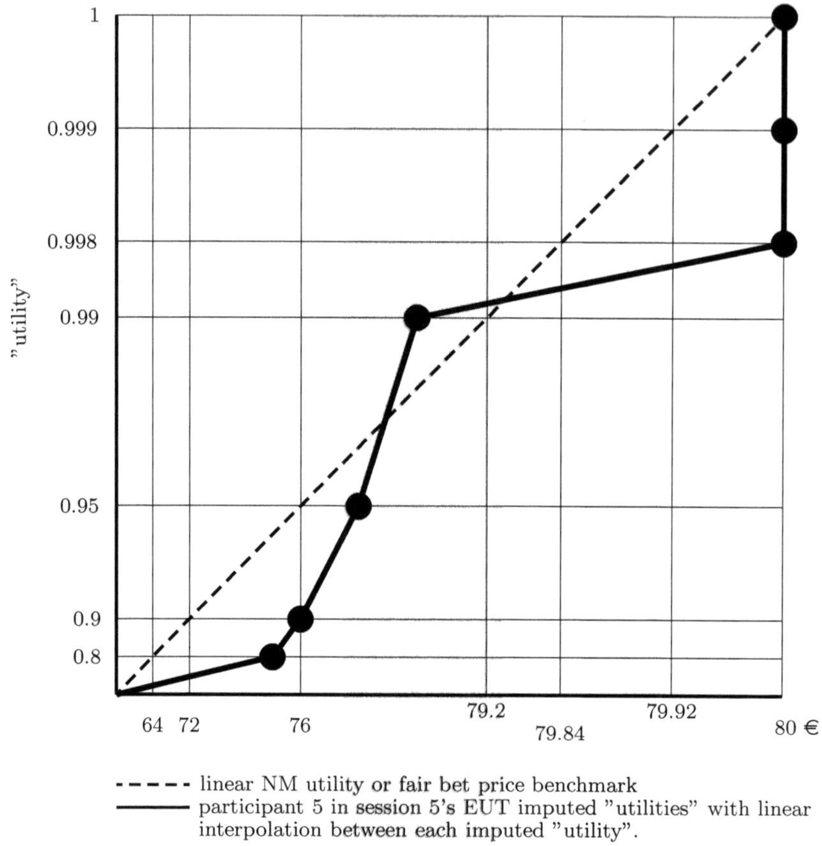

Risk of attack	**0.2**	**0.1**	**0.05**	**0.01**	**0.002**	**0.001**
x denotes chose to insure						
WTP in €	5	4	3	2	0	0
Compared to linear NM utility or fair bet	<	<	<	>	<	<
Certainty equivalent: 80 - WTP	75	76	77	78	80	80
Classification	Convex then Concave then Convex-Vertical Consistency of WTP and insurance decisions					

1 The scales on the axes are non-linear, but chosen such that (subject to graph grid limits), the fair bet choice benchmark denoting linear utility under EUT, is linear

2 The vertical axis shows the probability of obtaining €80 without insurance. The thick black line shows the connection of this probablity with a certainty equivalent amount, namely €80 minus the willingness to pay for insurance at this probability, with linear interpolations between adjacent amounts (circles). Under the assumption that the participant obeys the axioms of expected utility theory, one can normalize the utilities of €0 and €80 as 0 and 1 and the vertical axis represents the utilities and the thick black line denotes the utility function of the participant. But since the thick black line contains a vertical section, the participant does not obey the axioms of expected utility theory.

Figure 38: "Utilities" deduced from EUT for participant 6 in session 5.

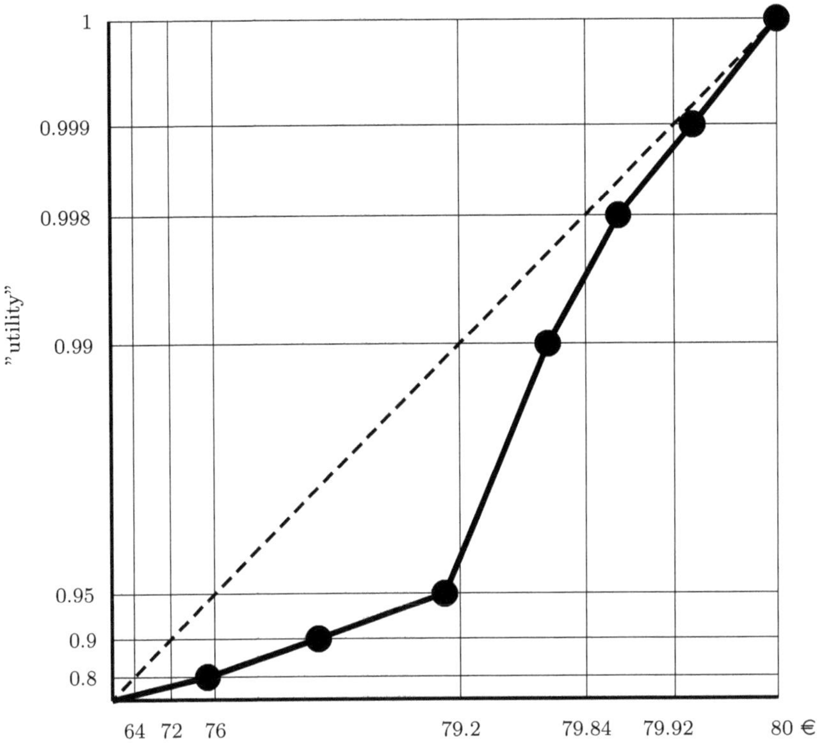

- - - - - linear NM utility or fair bet price benchmark
——— participant 6 in session 5's EUT imputed "utilities" with linear interpolation between each imputed "utility".

Risk of attack	**0.2**	**0.1**	**0.05**	**0.01**	**0.002**	**0.001**
x denotes chose to insure						
WTP in €	5	2.5	1.25	0.5	0.1	0.05
Compared to linear NM utility or fair bet	<	<	<	<	<	<
Certainty equivalent: 80 - WTP	75	77.5	78.75	79.5	79.9	79.95
Classification	Convex then Linear then Convex then Concave then Linear Consistency of WTP and insurance decisions					

1 The scales on the axes are non-linear, but chosen such that (subject to graph grid limits), the fair bet choice benchmark denoting linear utility under EUT, is linear

2 The vertical axis shows the probability of obtaining €80 without insurance. The thick black line shows the connection of this probablity with a certainty equivalent amount, namely €80 minus the willingness to pay for insurance at this probability, with linear interpolations between adjacent amounts (circles). Under the assumption that the participant obeys the axioms of expected utility theory, one can normalize the utilities of €0 and €80 as 0 and 1 and the vertical axis represents the utilities and the thick black line denotes the utility function of the participant.

Figure 39: "Utilities" deduced from EUT for participant 8 in session 5.

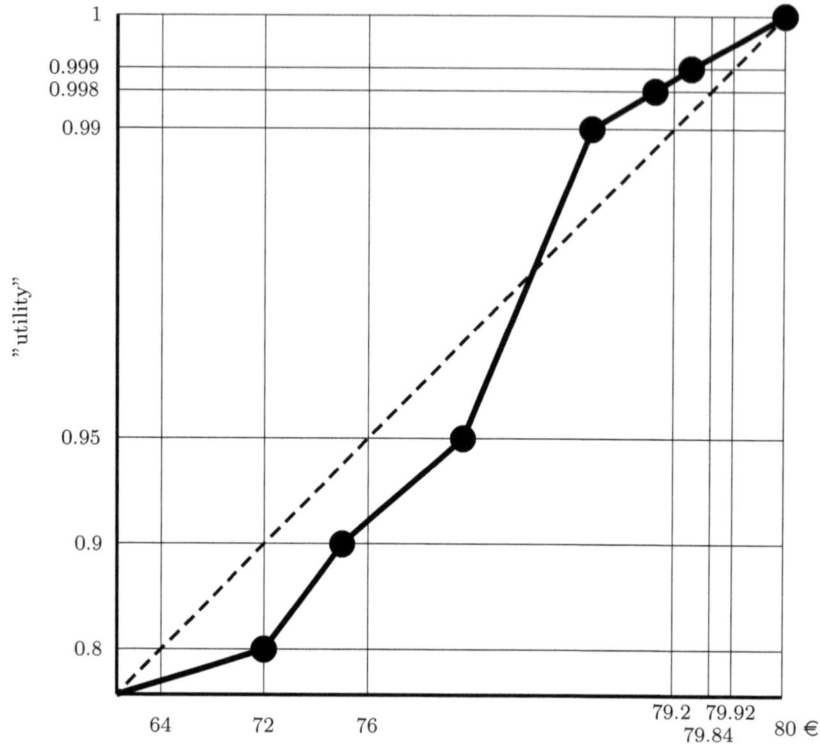

- - - - - linear NM utility or fair bet price benchmark
——— participant 8 in session 5's EUT imputed "utilities" with linear interpolation between each imputed "utility".

Risk of attack	0.2	0.1	0.05	0.01	0.002	0.001
x denotes chose to insure						
WTP in €	8	5	3	2	1	0.5
Compared to linear NM utility or fair bet	<	<	<	>	>	>
Certainty equivalent: 80 - WTP	72	75	77	78	79	79.5
Classification	Convex then Concave then Convex then Concave then Linear Consistency of WTP and insurance decisions					

1 The scales on the axes are non-linear, but chosen such that (subject to graph grid limits), the fair bet choice benchmark denoting linear utility under EUT, is linear

2 The vertical axis shows the probability of obtaining €80 without insurance. The thick black line shows the connection of this probablity with a certainty equivalent amount, namely €80 minus the willingness to pay for insurance at this probability, with linear interpolations between adjacent amounts (circles). Under the assumption that the participant obeys the axioms of expected utility theory, one can normalize the utilities of €0 and €80 as 0 and 1 and the vertical axis represents the utilities and the thick black line denotes the utility function of the participant.

Figure 40: "Utilities" deduced from EUT for participant 9 in session 5.

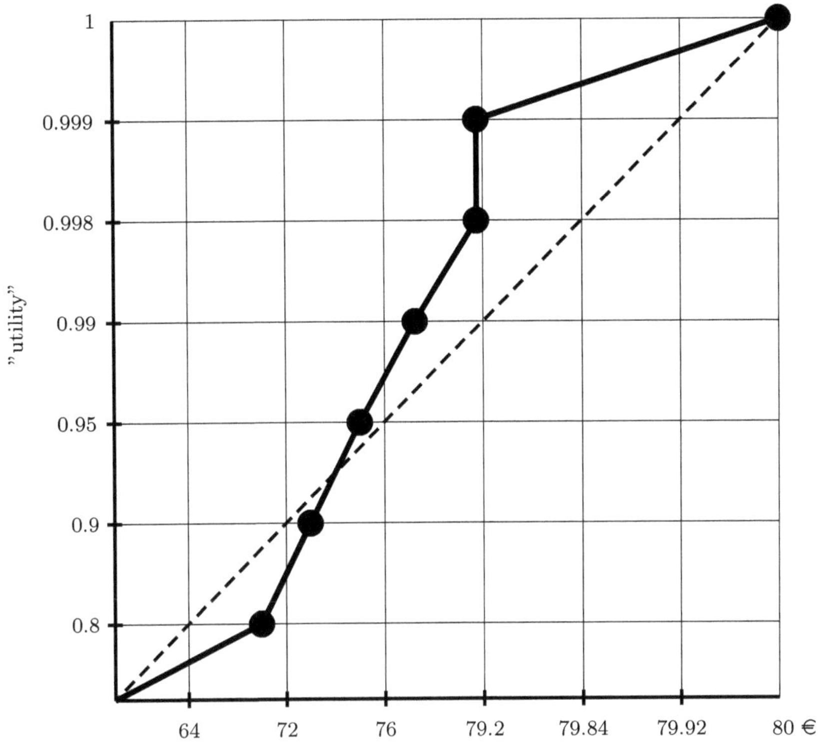

- - - - - linear NM utility or fair bet price benchmark
——— participant 9 in session 5's EUT imputed "utilities" with linear interpolation between each imputed "utility".

Risk of attack	**0.2**	**0.1**	**0.05**	**0.01**	**0.002**	**0.001**
x denotes chose to insure						
WTP in €	10	7	5	3	1	1
Compared to linear NM utility or fair bet	<	<	>	>	>	>
Certainty equivalent: 80 - WTP	70	73	75	77	79	79
Classification	Convex then Concave then Convex-Vertical then Concave Consistency of WTP and insurance decisions					

1 The scales on the axes are non-linear, but chosen such that (subject to graph grid limits), the fair bet choice benchmark denoting linear utility under EUT, is linear

2 The vertical axis shows the probability of obtaining €80 without insurance. The thick black line shows the connection of this probablity with a certainty equivalent amount, namely €80 minus the willingness to pay for insurance at this probability, with linear interpolations between adjacent amounts (circles). Under the assumption that the participant obeys the axioms of expected utility theory, one can normalize the utilities of €0 and €80 as 0 and 1 and the vertical axis represents the utilities and the thick black line denotes the utility function of the participant. But since the thick black line contains a vertical section, the participant does not obey the axioms of expected utility theory.

Figure 41: "Utilities" deduced from EUT for participant 11 in session 5.

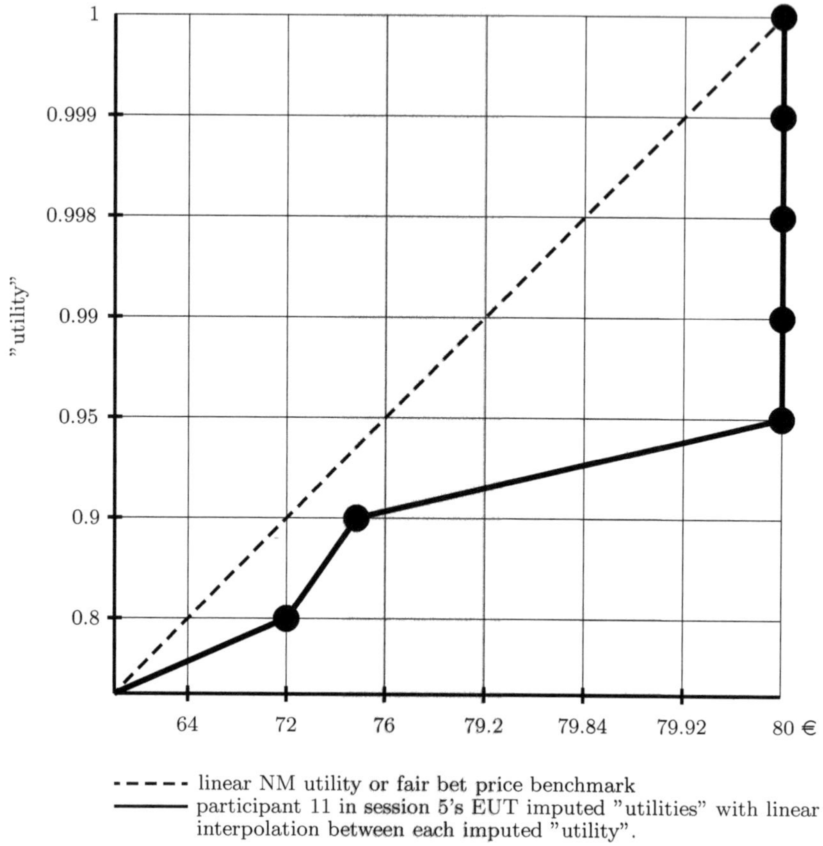

Risk of attack	0.2	0.1	0.05	0.01	0.002	0.001
x denotes chose to insure						
WTP in €	8	5	0	0	0	0
Compared to linear NM utility or fair bet	<	<	<	<	<	<
Certainty equivalent: 80 - WTP	72	75	80	80	80	80
Classification	Convex then Concave then Convex-Vertical Consistency of WTP and insurance decisions					

1 The scales on the axes are non-linear, but chosen such that (subject to graph grid limits), the fair bet choice benchmark denoting linear utility under EUT, is linear

2 The vertical axis shows the probability of obtaining €80 without insurance. The thick black line shows the connection of this probablity with a certainty equivalent amount, namely €80 minus the willingness to pay for insurance at this probability, with linear interpolations between adjacent amounts (circles). Under the assumption that the participant obeys the axioms of expected utility theory, one can normalize the utilities of €0 and €80 as 0 and 1 and the vertical axis represents the utilities and the thick black line denotes the utility function of the participant. But since the thick black line contains a vertical section, the participant does not obey the axioms of expected utility theory.

Figure 42: "Utilities" deduced from EUT for participant 12 in session 5.

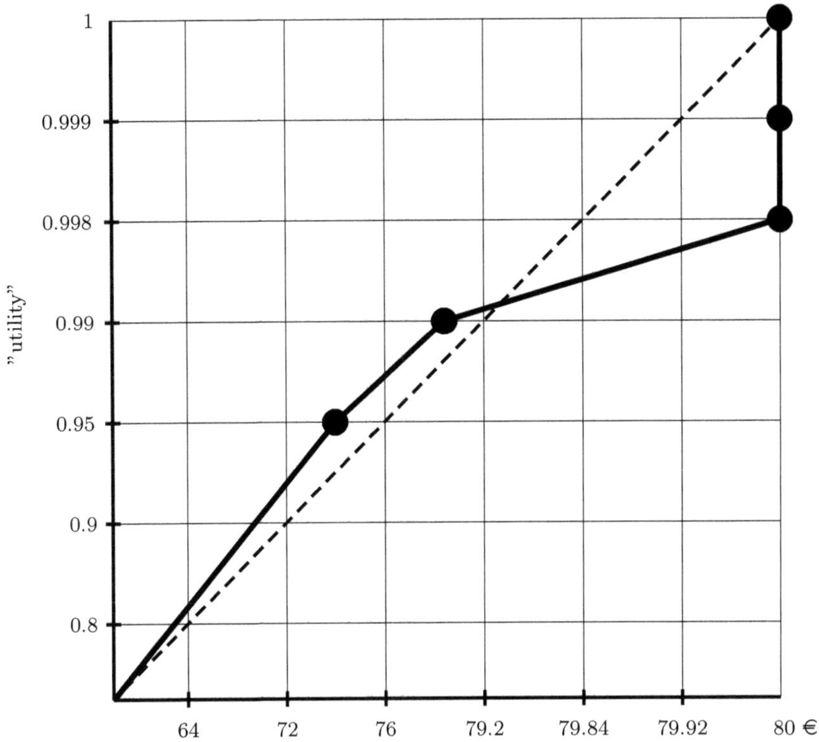

- - - - linear NM utility or fair bet price benchmark
——— participant 12 in session 5's EUT imputed "utilities" with linear interpolation between each imputed "utility".

Risk of attack	0.2	0.1	0.05	0.01	0.002	0.001
x denotes chose to insure	x	x				
WTP in €	E	E	6	2	0	0
Compared to linear NM utility or fair bet	E	E	>	>	<	<
Certainty equivalent: 80 - WTP	E	E	74	78	80	80
Classification	Concave then Convex-Vertical Consistency of WTP and insurance decisions					

E denotes that the participants left these fields empty in the questionnaire

1 The scales on the axes are non-linear, but chosen such that (subject to graph grid limits), the fair bet choice benchmark denoting linear utility under EUT, is linear

2 The vertical axis shows the probability of obtaining €80 without insurance. The thick black line shows the connection of this probablity with a certainty equivalent amount, namely €80 minus the willingness to pay for insurance at this probability, with linear interpolations between adjacent amounts (circles). Under the assumption that the participant obeys the axioms of expected utility theory, one can normalize the utilities of €0 and €80 as 0 and 1 and the vertical axis represents the utilities and the thick black line denotes the utility function of the participant. But since the thick black line contains a vertical section, the participant does not obey the axioms of expected utility theory.

Figure 43: "Utilities" deduced from EUT for participant 13 in session 5.

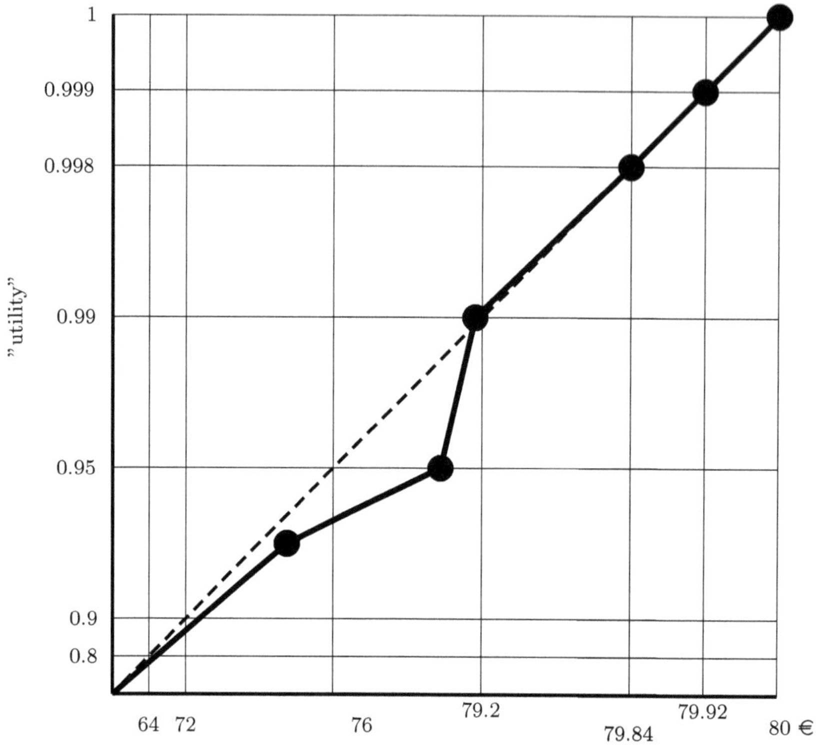

- - - - - linear NM utility or fair bet price benchmark
——— participant 13 in session 5's EUT imputed "utilities" with linear interpolation between each imputed "utility".

Risk of attack	**0.2**	**0.1**	**0.05**	**0.01**	**0.002**	**0.001**
x denotes chose to insure	x					
WTP in €	E	7	2	1	0.2	0.1
Compared to linear NM utility or fair bet	E	<	<	>	>	>
Certainty equivalent: 80 - WTP	E	73	78	79	79.8	79.9
Classification	Concave then Convex then Concave then Linear Consistency of WTP and insurance decisions					

E denotes that the participants left these fields empty in the questionnaire

1 The scales on the axes are non-linear, but chosen such that (subject to graph grid limits), the fair bet choice benchmark denoting linear utility under EUT, is linear
2 The vertical axis shows the probability of obtaining €80 without insurance. The thick black line shows the connection of this probablity with a certainty equivalent amount, namely €80 minus the willingness to pay for insurance at this probability, with linear interpolations between adjacent amounts (circles). Under the assumption that the participant obeys the axioms of expected utility theory, one can normalize the utilities of €0 and €80 as 0 and 1 and the vertical axis represents the utilities and the thick black line denotes the utility function of the participant.

Figure 44: "Utilities" deduced from EUT for participant 14 in session 5.

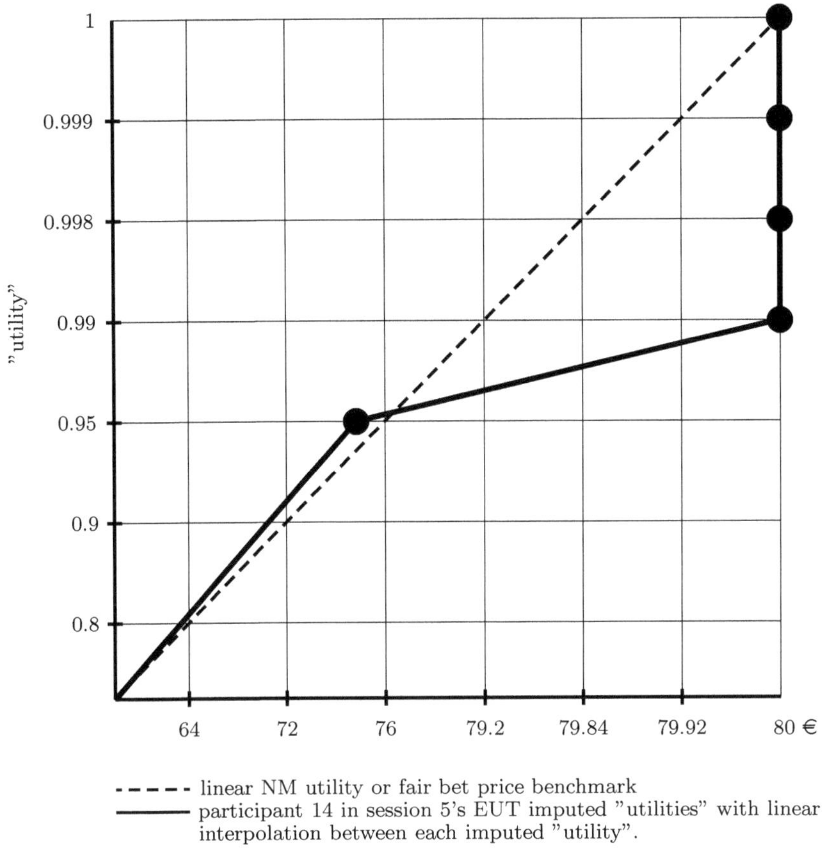

Risk of attack	0.2	0.1	0.05	0.01	0.002	0.001
x denotes chose to insure	x	x				
WTP in €	E	E	5	0	0	0
Compared to linear NM utility or fair bet	E	E	>	<	<	<
Certainty equivalent: 80 - WTP	E	E	75	80	80	80
Classification	Concave then Convex-Vertical Consistency of WTP and insurance decisions					

E denotes that the participants left these fields empty in the questionnaire

1 The scales on the axes are non-linear, but chosen such that (subject to graph grid limits), the fair bet choice benchmark denoting linear utility under EUT, is linear
2 The vertical axis shows the probability of obtaining €80 without insurance. The thick black line shows the connection of this probablity with a certainty equivalent amount, namely €80 minus the willingness to pay for insurance at this probability, with linear interpolations between adjacent amounts (circles). Under the assumption that the participant obeys the axioms of expected utility theory, one can normalize the utilities of €0 and €80 as 0 and 1 and the vertical axis represents the utilities and the thick black line denotes the utility function of the participant. But since the thick black line contains a vertical section, the participant does not obey the axioms of expected utility theory.

Appendix C

Questionnaire

These sheets were given Enrolees on Entering Experiment Advertised as on Behaviour Under Terrorism. Darden Graduate School of Business Administration, University of Virginia, 16 March 04.

One of you will be selected by random draw to gain \$100 tomorrow afternoon from me except that there could be an attack in the meantime, with the attackers taking the \$100. Should an attack take place, you gain nothing unless you paid an insurance fee of \$15 to ensure that your \$100 was protected overnight from the attack.
We shall invite each of you to write your name on a piece of paper, and put these in a box and shuffle them. Then we shall invite one of you to draw out a paper from the box to determine who is the potential recipient of the \$100
Before doing this draw to select one of you for the \$100, we ask each of you to indicate the circumstances under which you would pay the \$15 protection money and thus receive \$100 - \$15 = \$85 for sure. We ask you to indicate whether you would pay under the following 6 different circumstances of how frequently in the last 1000 days there has been an attack.

200	100	50	10	2	1

Please put a cross in those frequencies for which you would pay the protection money. We shall collect your decisions and your decision will be binding in the event that you are selected as the potential recipient

of the $100.
Note that these numbers of days correspond to the below percentages of days on which there was an attack in the last 1000 days.

20%	10%	5%	1%	0.2%	0.1%

The participant selected today as the potential recipient of the $100 will write each of the 6 numbers 200, 100, 50, 10, 2 and 1 on a separate piece of paper placed in a bag. The papers will then be shuffled and the selected participant will then draw one out. This determines the actual likelihood of an attack in the next 24 hours, where likelihood is determined by marked matches in a bag as follows.
Tomorrow afternoon, the potential recipient receives the $85 if that person had chosen paid the $15 protection money for the actual likelihood of an attack. Otherwise the potential recipient draws a match out of a box with a proportion of marked matches, the proportion corresponding to the actual likelihood of an attack. If the potential recipient draws out a marked match, that person receives nothing, if an unmarked match, $100.
Before we proceed to determine the potential recipient of the $100, please fill out the following questions on why you did or did not decide to pay the protection money for each of the 6 likelihoods of attack.
Question Set A In those circumstances in which you chose to pay for protection, which considerations led you to this decision:

1. To avoid worry that in an attack I would lose it all and/or
2. $85 is far better than $0. But $100 is only a little better than $85 and/or
3. Other reasons. Please explain these.

Question Set B In those circumstances in which you chose to not pay for protection, which considerations led you to this decision:

1. To enjoy the thrill that there might be an attack where I would lose all and/or
2. $85 is only slightly better than $0. But $100 is far better than $85 and/or
3. Other reasons. Please explain these.

Appendix D

Glossary / Acronyms

Dominance Principle: A preference for FOSD, first order stochastically dominant distributions over outcomes. This principle excludes consideration of secondary satisfactions, limits the decision maker to considering only primary satisfactions.

Elaborated or re-defined outcomes: a conjecture of Markowitz (1959) and Samuelson (1952a) that EUT could include emotional secondary satisfactions by elaborating or redefining the outcomes space to include secondary satisfactions, a conjecture proven in Pope (1991b) to destroy the axiomatic base of EUT and in Pope (2000) to deprive EUT of a representation theorem in the normal sense.

FOSD: a First Order Stochastically Dominant distributions over outcomes. Examples of FOSD are provided in chapters

EUT: axiomatised Expected Utility theory

EUT+: any decision theory that imposes the dominance principle

Primary satisfactions: positive or negative satisfactions reaped independent of knowledge ahead. For primary satisfactions, it is irrelevant to the chooser what is his degree of knowledge ahead at any stage, from the stage when a problem is first to discerned, through the stage when a decision is taken, to the final stage beginning when he learns the outcome of his decision.

Post-outcome period: the period that begins when the chooser learns the outcome of his chosen act, and thus his risk his passed, his entire future with respect to that outcome is certain - he has in this regard full knowledge ahead.

Pre-outcome period: the period that begins on choosing an act and lasts until the chooser learns the outcome of his act, ie the period during which he has risk, uncertainty. A limited degree of knowledge ahead of the outcome of his act.

Process utility: the term used in Harsanyi (1978) for (emotional) secondary satisfactions.

Rank Dependent theory: a theory in which the chooser globally maximizes by selecting an act in his choice set whose value (number) is a maximum. In turn he determines the value (number) of each act by aggregating the mutually exclusive utilities / payoffs either by simple probability weights, or by functions of the (de)cumulative distribution. Examples of rank dependent theories include expected (discounted) net revenues, EUT, Quiggin's anticipated utility theory (Quiggin (1982, 1993)), Allais' Invariance Theory (Allais (1988)), and Tversky and Kahneman's cumulative prospect theory (Tversky & Kahneman (1992)). All these are standard rank dependent theories in the sense that they impose the dominance principle, ie exclude secondary satisfactions in how outcomes map into utilities.

Risk attitude: a term used in Luce & Raiffa (1957) for what are in fact secondary satisfactions. But since by 1957 the words "risk aversion" had already been changed to have an EUT meaning, there has also evolved a tradition of using risk attitude to denote the shape of an EUT adherent's "as if certain" utility mapping. Many economists inadvertently slip between meaning this shape and meaning secondary satisfactions by risk attitude, not realising that these are distinct concepts.

Risk aversion: a term that prior to Marschak (1950) denoted a negative secondary satisfaction, namely a dislike of dispersion in outcomes.

After 1950 within EUT these words have denoted a concave "as if certain" mapping from outcomes into utilities, as in the Arrow-Pratt risk aversion measures. The words "risk aversion" continue to have their original denotation in that branch of the finance literature that employs models such as CAPM that are in general incompatible with EUT. Many economists inadvertently slip between the two meanings of the words. This robs them of understanding of what EUT includes, and what EUT excludes.

Risk loving: a term whose meaning was changed in Marschak (1950) to mean an "as if certain" EUT mapping from money outcomes into utilities that is convex

Risk neutrality: a term whose meaning was changed in Marschak (1950) to mean an "as if certain" EUT mapping from money outcomes into utilities that is linear

Secondary satisfactions: positive or negative satisfactions affected by the chooser's degree of knowledge ahead at any stage. Secondary satisfactions occur whenever the chooser is affected by his degree of knowledge ahead at any stage, from the stage when a problem is first to discerned, through the stage when a decision is taken, to after the outcome of the decision is learned.

SKAT: Stages of Knowledge Ahead Theory. This theory divides the chooser's future by his degree of knowledge ahead. Initially on perceiving a situation, he does not know his alternatives. Later he has a change in knowledge ahead, having discovered at least two alternative acts (one of which may be doing nothing). Still he does not know which act he will choose. Later he has a change in knowledge ahead, having chosen an act. Still if the act is risky, he does not know the outcome of his act. He is in his pre-outcome period. Later he has a change in knowledge ahead, having learned the outcome of his chosen act. He is in his post-outcome period. This division of the future by evolving stages of knowledge ahead allows the chooser to anticipate his secondary satisfactions, that is his satisfactions based on his degree of knowledge ahead at each stage, and have these affect his choice of an act.

Utility of gambling: a term sometimes prefaced with specific, denoting emotional secondary satisfactions. The pejorative, frivolous even irrational connotations of the term stem from a focus on social games of chance, not serious decisions.

References

Abbink, K., Sadrieh, A., & Selten, R. (1999). Money does not induce risk neutral behavior, but binary lotteries do even worse. *Theory and Decision, 46*, 213-252.

Albers, W. (2001). Prominence theory as a tool to model boundedly rational decisions. In G. Gigerenzer & R. Selten (Eds.), *Bounded rationality: The adaptive toolbox.* MIT Press, Cambridge Massachusetts.

Albers, W., Pope, R., Selten, R., & Vogt, B. (2000). Experimental evidence for attractions to chance. *German Economic Review, 1*, 113-129.

Alchian, A. (1950). Uncertainty, evolution, and political economy. *Journal of Political Economy, 58*, 211-231.

Allais, M. (1952). Fondements d'une théorie positive des choix comportant un risque et critique des postulats et axiomes de l'ecole américaine (the foundations of a positive theory of choice involving risk and a criticism of the postulates and axioms of the american school). In M. Allais & O. Hagen (Eds.), *Expected utility hypotheses and the allais paradox: Contemporary discussions of decisions under uncertainty with allais' rejoinder.* Reidel, Dordrecht.

Allais, M. (1988). The general theory of random choices in relation to the invariant cardinal utility function and the specific probability function. the (u, q) model: A general overview. In B. Munier (Ed.), *Risk, decision and rationality* (p. 231-289). Reidel, Dordrecht.

Barker, D. (2003). Terrorism insurance subsidies and social welfare. *Journal of Urban Economics, 54*, 328-338.

Barnett, S., & Cowan, P. (1976). Activity, exploration, curiosity and fear: An ethological study. *Interdisciplinary Science Reviews, 1*, 43-62.

Bell, D. (1981). Components of risk aversion. In J. Brans (Ed.), *Operational research* (p. 371-378). North-Holland, Amsterdam.

Bell, D. (1985). Disappointment in decision making under uncertainty. *Operations Research, 33*, 1-27.

Berg, C. (1970). Drei Oligopolspiele in kontinuierlicher Zeit. In H. Sauermann (Ed.), *Beiträge zur experimentellen Wirtschaftsforschung* (Vol. 2). Tübingen.

Bernard, G. (1966). Sur les fonctions d'utilité. *Revue Francaise de Recherche Opérationnelle, 41*, 323-352.

Bernard, G. (1970). *On utility functions.* Centre d'Etude des Techniques Economiques Modernes, Paris.

Bernholz, P. (2004). Supreme values as the basis for terror. *European Journal of Political Economy, 20*, 317-333.

Berns, G., McClure, S., Pagnoni, G., & Montague, P. (2001). Predictability modulates human brain response to reward. *Journal of Neuroscience, 21*, 2793-2798.

Black, M. (1986). Some questions about bayesian decision theory. In L. Daboni, A. Montesano, & M. Lines (Eds.), *Recent developments in the foundations of utility and risk theory* (p. 57-66). Reidel, Dordrecht.

Bleichrodt, H., Pinto, J., & Wakker, P. (2001). Making descriptive use of prospect theory to improve the prescriptive use of expected utility. *Management Science, 47*, 1498-1514.

Blomberg, S., Hess, G., & Weerapana, A. (2003). Economic conditions and terrorism. *European Journal of Political Economy, 20*, 463-478.

Borch, K. (1969). A note on uncertainty and indifference curves. *Review of Economic Studies, 36*, 1-4.

Borcherding, K., Eppel, T., & Winterfeldt, D. von. (1991). Comparison of weighting judgments in multiattribute utility measurement. *Management Science, 37*, 1603-1619.

Borcherding, K., Schmeer, S., & Weber, M. (1995). Biases in multiattribute weight elicitation. In J.-P. Caverni, H. M. Bar-Hillel, H. Barrona, H. Jungermann, & O. Huber (Eds.), *Contributions to decision making I* (p. 3-28). North-Holland.

Borcherding, K., & Winterfeldt, D. von. (1988). The effect of varying value trees on multiattribute evaluations. *Acta Psychologica, 68*, 153-170.

Bosman, R., & Winden, F. van. (2005). *Global risk, investment and emotions.* (Working paper, CREED, The Center for Research in Experimental Economics and Political Decision-making is a research Institute at the Faculty of Economics and Econometrics, University of Amsterdam.)

Brandstätter, E., Gigerenzer, G., & Hertwig, R. (2006). The priority heuristic. making choices without trade-offs. *Psychological Review, 113*, 409-432.

Brown, T. (forthcoming). Loss aversion without the endowment effect, and other explanations for the WTA-WTP disparity. *Journal of Economic Behavior and Organization.*

Brück, T., & Wickström, B.-A. (2004). The economic consequences of terror: guest editors' introduction. *European Journal of Political Economy, 20*, 293-300.

Camille, N., Coricelli, G., Sallet, J., Pradat-Diehl, P., Duhamel, J.-R., & Sirigu, A. (2004). The involvement of the orbitofrontal cortex in the experience of regret. *Science, 21*, 1167-1170.

Canaan, E. (1926, reprinted 1963). Profit. In H. Higgs (Ed.), *Palgrave's dictionary of political economy* (p. 222-224). Augustus M. Kelley, New York.

Caplin, A., & Leahy, J. (2001). Psychological expected utility theory and anticipatory feelings. *The Quarterly Journal of Economics, 1*, 55-79.

Chen, A. (2006). *Loss analysis of a life insurance company applying discrete-time risk-minimizing hedging strategies.* (Bonn Graduate School of Economics Working Paper No. 19.)

Chen, A., & Siems, T. (2004). The effects of terrorism on global capital markets. *European Journal of Political Economy, 20*, 349-366.

Conlisk, J. (1993). The utility of gambling. *Journal of Risk and Uncertainty, 6*, 255-275.

Coombs, C., & Huang, L. (1970). Tests of a portfolio theory of risk preference. *Journal of Experimental Psychology, 85*, 23-29.

Coursey, D. L., Hovis, J. L., & Schulze, W. D. (1987). The disparity between willingness to accept and willingness to pay measures of value. *Quarterly Journal of Economics, 102*(3), 679-690.

Cyert, R., & March, J. (1963). *The behavioral theory of the firm.* Prentice Hall.

Damasio, A. (1984). *Emotion, reason and the human brain.* Avon Book, New York.

Darwin, C. (1874). *The descent of man.* Murray, London.

Del Mar, C. (2002). Should we debrief and counsel people who have had a psychological shock? *The Medical Journal of Australia, 177*, 258-260.

De Neufville, R., & Delquie, P. (1988). A model of the influence of certainty and probability "effects" on the measurement of utility. In B. Munier (Ed.), *Risk, decision and rationality* (p. 189-205). Reidel, Dordrecht.

De Neufville, R., & McCord, M. (1983a). A decision analysis model when the substitution principle is not acceptable. In B. Stigum & F. Wenstop (Eds.), *Foundations of utility and risk theory with applications* (p. 247-262). D. Reidel, Holland.

De Neufville, R., & McCord, M. (1983b). Empirical demonstration that expected utility decision analysis is not optimal. In B. Stigum & F. Wenstop (Eds.), *Foundations of utility and risk theory with applications* (p. 181-199). D. Reidel, Holland.

De Neufville, R., & Odoni, A. (2002). *Airport systems planning, design and management.* McGraw-Hill, New York.

Diecidue, E., Schmidt, U., & Wakker, P. (1999). *A theory of the gambling effect.* (mimeo, CentER, Tilburg University, Tilburg)

Dohmen, T., Falk, A., Huffman, D., Sunde, U., Schupp, J., & Wagner, G. (2005). *Individual risk attitudes: New evidence from a large, representative, experimentally-validated survey.* (IZA Discussion Paper No. 1730)

Dreze, J., & Modigliani, F. (1966). Epargne et consommation en avenir aleatoire. *Cahiers du Seminaire d'Econometrie, 9*, 7-33.

Dreze, J., & Modigliani, F. (1972). Consumption decisions under uncertainty. *Journal of Economic Theory, 5*, 308-355.

Dufwenberg, M. (2002). Marital investment, time consistency, and emotions. *Journal of Economic Behavior and Organization, 48*, 57-69.

Eden, B. (1977). The role of insurance and gambling in allocating risk over time. *Journal of Economic Theory, 5*, 308-335.

Ellsberg, D. (1954). Classic and current notions of "measurable utility". *Economic Journal, 64*, 528-556.

Feldstein, M. (1969). Mean-variance analysis in the theory of liquidity preference and portfolio selection. *Review of Economic Studies, 36*, 5-12.

Fitzgerald, V. (2004). Global financial information, compliance incentives and terrorist funding. *European Journal of Political Economy, 20*, 387-401.

Foster, J. (2006). Letter: The Costs to the Iraqi People May Become Costs to the United States. *The Economists' Voice, 3*. (http://www.bepress.com/ev/vol3/iss4/art7)

Fredrickson, B., & Kahneman, D. (1993). Duration neglect in retrospective evaluations of affective episodes. *Journal of Personality and Social Psychology, 65*, 45-55.

Frey, B. S., & Luechinger, S. (2004). Decentralization as a disincentive for terror. *European Journal of Political Economy, 20*, 509-515.

Friedman, M., & Savage, L. (1948). The utility analysis of choices involving risk. *Journal of Political Economy, 56*, 279-304.

Gale, D., & Hellwig, M. (1985). Incentive-compatible debt contracts: One period problem. *Review of Economic Studies, 52*, 647-663.

Gigerenzer, G. (2003). The adaptive toolbox and life span development: Common questions? In U. Staudinger & U. Lindenberger (Eds.), *Understanding human development: Life span psychology in exchange with other disciplines* (p. 423-435). Dordrecht, Kluwer.

Gigerenzer, G. (2004). Striking a blow for sanity in theories of rationality. In M. Augier & M. J.G. (Eds.), *Models of a Man: Essays in the Memory of Herbert A. Simon* (p. 389-409). Cambridge, MA.

Gigerenzer, G., & Todd, P. (forthcoming). Rationality the fast and frugal way: Introduction. In C. R. Plott & V. L. Smith (Eds.), *Handbook of experimental economics results.* Amsterdam, Netherlands: Elsevier/North-Holland.

Gilboa, I., & Schmeidler, D. (2001). *A theory of case-based decisions.* Cambridge University Press, Cambridge UK.

Grant, S., A., K., & Polak, B. (1998). Intrinsic preference for information. *Journal of Economic Theory, 83*, 233-259.

Gul, F. (1991). A theory of disappointment aversion. *Econometrica, 59*, 667-686.

Hagen, O. (1985). Rules of behaviour and expected utility theory - compatibility versus dependence. *Theory and Decision, 18*, 31-45.

Hardie, E. (2002). *A tripartite model of self, stress and uplifts, coping and health.* (paper presented to New Zealand Health Psychology and Behavioural Medicine Conference, Auckland New Zealand.)

Harless, D. W. (1989). More laboratory evidence on the disparity between willingness to pay and compensation demanded. *Journal of Economic Behavior and Organization, 11*, 359-379.

Harrison, G., & List, J. (2004). Field experiments. *Journal of Economic Literature, 42*, 1013-1059.

Harsanyi, J. (1978). Bayesian decision theory and utilitarian ethics. *American Economic Review, 68*, 223-228.

Hart, A. (1930). *Anticipations, uncertainty and dynamic planning.* Augustus M. Keeley, Inc., New York.

Hassenzahl, M., & Borcherding, K. (2004). *The endowment effect for hedonic and utilitarian objects.* (Tagung des sozialwissenschaftlichen Ausschuss des Verein für Socialpolitik, 22.-24. April 2004, Graz)

Hausken, K. (2004). *Quantifying the relationship between risk and price.* (mimeo, Department of Social Sciences, University of Stavanger, Norway)

Hebb, D. (1949). *The organization of behavior: A neuropsychological theory.* John Wiley & Sons, Inc.

Hershey, J. C., Kunreuther, H., & Schoemaker, P. (1982). Sources of bias in assessment procedures for utility functions. *Management Science, 28*, 936-954.

Hershey, J. C., & Schoemaker, P. J. H. (1985). Probability vs. certainty equivalence methods in utility measurement: Are they equivalent? *Management Science, 31*, 1213-1231.

Hey, J. (2005). Comparing theories: What are we looking for? In U. Schmidt & S. Traub (Eds.), *Advances in Public Economics: Utility, Choice and Welfare - A Festschrift for Christian Seidl.* Kluwer.

Hey, J., & Orme, C. (1994). Investigating generalizations of expected utility theory using experimental data. *Econometrica, 62*, 1291-1326.

Hoel, M., Iversen, T., Nilssen, T., & Vislie, J. (forthcoming). Genetic testing in competitive insurance markets with repulsion from chance: a welfare analysis. *Journal of Health Economics.*

Hogarth, R. (2006). *A new paradigm, presentation to the roundtable.* (Behavioral Economics and Economic Psychology Conference, Paris 5-8 July)

Jaffee, D., & Russell, T. (2006). Should governments provide catastrophe insurance? *The Economists' Voice, 3,* Article 6. (http://www.bepress.com/ev/vol3/iss5/art6)

Jain, S., & Mukand, S. W. (2004). The economics of high-visibility terrorism. *European Journal of Political Economy, 20,* 479-494.

Janis, I., & Mann, L. (1977). *Decision making: A psychological analysis of conflict, choice, and commitment.* Free Press, New York.

Kahneman, D. (2000). Experienced utility and objective happiness: A moment-based approach. In D. Kahneman & A. Tversky (Eds.), *Choices, values and frames* (p. 673-692). Cambridge University Press and the Russell Sage Foundation.

Kahneman, D., Fredrickson, D., Schreiber, C., & Redelemeier, D. (1993). When more pain is preferred to less: Adding a better end. *Psychological Science, 4,* 401-405.

Kahneman, D., & Tversky, A. (1979). Prospect theory: An analysis of decision under risk. *Econometrica, 47,* 263-291.

Karni, E. (2006). Subjective expected utility theory without states of the world. *Journal of Mathematical Economics, 42,* 325-342.

Kiyotaki, N., & Moore, J. (1997). Credit cycles. *Journal of Political Economy, 105,* 211-248.

Kliemt, H. (2004). *Simulation, stimulation, stipulation - reflections on the means and ends of experimental economics.* (Talk given on the GEW-Tagung in Prague, October 15, 2004)

Knight, F. (1933). *Risk, uncertainty, and profit.* Series of Reprints of Scarce Tracts in Economics and Political Science, no. 16. London: London School of Economics and Political Science.

Kreps, D., & Porteus, E. (1978). Temporal resolution of uncertainty and dynamic choice theory. *Econometrica, 46,* 185-200.

Kreps, D., & Porteus, E. (1979a). Dynamic choice theory and dynamic programming. *Econometrica, 47*, 91-100.

Kreps, D., & Porteus, E. (1979b). Temporal von Neumann-Morgenstern and Induced Preferences. *Journal of Economic Theory, 20*, 81-109.

Krzysztofowicz, R. (1983). Strength of preference and risk attitude in utility measurement. *Organizational Behavior and Human Performance, 31*, 88-113.

Kunreuther, H. (2002). The role of insurance in managing extreme events: Implications for terrorism coverage. *Business Economics, 22*, 6-16.

Lakdawalla, D., & Zanjani, G. (2005). Insurance, self-protection, and the economics of terrorism. *Journal of Public Economics, 89*, 1891-1905.

Langlois, R., & Cosgel, M. (1993). Frank Knight on Risk, Uncertainty and the Firm: A New Interpretation. *Economic Inquiry, 31*, 456-465.

Lazarus, R. (2000). Toward better research on stress and coping. *American Psychologist, 55*, 665-673.

Loewenstein, G. (1987). Anticipation and the value of delayed consumption. *Economic Journal, 97*, 666-684.

Loewenstein, G., & Adler, D. (1995). A bias in the prediction of tastes. *Economic Journal, 105*, 929-937.

Loomes, G., & Sugden, R. (1982). Regret theory: An alternative theory of rational choice under uncertainty. *Economic Journal, 92*, 805-824.

Lopes, L. (1984). Risk and distributional inequality. *Journal of Experimental Psychology: Human Perception and Performance, 10*, 465-485.

Lopes, L. (1987). Between hope and fear: the psychology of risk. *Advances in Experimental Social Psychology, 20*, 255-295.

Lopes, L., & Oden, G. (1999). The role of aspiration level in risky choice: A comparison of cumulative prospect theory and sp/a theory. *Journal of Mathematical Psychology, 43*, 286-313.

Luce, R., & Marley, A. (forthcoming). Independence properties vis-a-vis several utility representations. *Theory and Decision.*

Luce, R., & Raiffa, H. (1957). *Games and decisions.* Wiley, New York.

Markowitz, H. (1959). *Portfolio selection: Efficient diversification of investments.* Wiley, New York.

Marschak, J. (1950). Rational behavior, uncertain prospects, and measurable utility. *Econometrica, 18*, 111-141.

Marschak, J., & Radner, R. (1972). *Economic theory of teams.* Yale University Press, New Haven.

McClennen, E. (1988). Dynamic choice and rationality. In B. Munier (Ed.), *Risk decision and rationality* (p. 517-536). Reidel, Dordrecht.

McClennen, E. (1990). *Rationality and dynamic choice, foundational explorations.* Cambridge University Press, Cambridge.

Mellers, B., A. Schwartz, K. H., & Ritov, I. (1997). Decision affect theory: Emotional reactions to the outcomes of risky options. *Psychological Science, 8*, 423-429.

Mellers, B., Erev, I., Fessler, D., Hemelrijk, C., Hertwig, R., Laland, K., et al. (2001). Group report: Effects of emotions and social processes on bounded rationality. In G. Gigerenzer & R. Selten (Eds.), *Bounded rationality the adaptive tool box.* MIT Cambridge Massachusetts.

Mellers, B., Schwartz, A., & Ritov, I. (1999). Emotion-based choice. *Journal of Experimental Psychology, 128*, 332-345.

Menstrel, L. (1999). *A process utility approach to the utility of gambling.* (presentation to Ninth International Conference on the Foundations and Applications of Utility, Risk and Decision Theory, Marrakesh)

Morgenstern, O. (1974). Some reflections on utility. In M. Allais & O. Hagen (Eds.), *Expected utility hypotheses and the allais paradox; contemporary discussions of decisions under uncertainty with allais' rejoinder* (p. 175-183). D. Reidel, Dordrecht.

Morgenstern, O. (1979). Some reflections on utility. In M. Allais & O.Hagen (Eds.), *Expected utility and the allais paradox* (p. 175-183). Dortrecht, Reidl.

Mossin, J. (1969). Communications: A note on uncertainty and preferences in a temporal context. *American Economic Review, 59*, 172-174.

Munera, H. (1978). *Modelling of individual risk attitudes in decision making under uncertainty: An application to nuclear power.* (Ph.D. dissertation, University of California, Berkeley)

Munier, B. (1996). Comment: Hammond's consequentialism: A qualification. In K. Arrow, M. Colombatto, M. Pearlman, & C. Schmidt (Eds.), *Rational foundations of economic behaviour.* McMillan, London.

Neumann, J. von, & Morgenstern, O. (1947). *Theory of games and economic behaviour* (Second ed.). Princeton University Press, Princeton.

Neumann, J. von, & Morgenstern, O. (1972). *Theory of games and economic behaviour* (Third ed.). Princeton University Press, Princeton.

Nitsch, V., & Schumacher, D. (2004). Terrorism and international trade: an empirical investigation. *European Journal of Political Economy*, *20*, 423-433.

Omodei, M., Wearing, A., McLennan, J., Elliott, G., & Clancy, J. (forthcoming). ore is better? a bias towards overuse of resources in naturalistic decision making settings. In R. Montgomery, R. Lipshitz, & B. Brehme (Eds.), *How professionals make decisions*. McMillan, London.

Pascal, B. (1670). *Pensées.* reprinted by Chez L Hachette, Paris.

Plato. (1967). Theaetetus. In F. M. Cornford (Ed.), *Plato's theory of knowledge: The theaetetus and the sophist, translated with a running commentary.* Routledge and Kegan Paul, London Rational.

Pope, R. (1983). The pre-outcome period and the utility of gambling. In B.P. Stigum & F. Wenstøp (Eds.), *Foundations of utility and risk theory with applications* (p. 37-177). D. Reidel, Dordrecht.

Pope, R. (1984). The utility of gambling and of outcomes: Inconsistent first approximations. In O. Hagen & F. Wenstøp (Eds.), *The utility of gambling and of outcomes: Inconsistent first approximations* (p. 251-273). D. Reidel, Dordrecht.

Pope, R. (1985). Timing Contradictions in von Neumann and Morgenstern's Axioms and in Savage's Sure-Thing Proof. *Theory and Decision*, *18*, 229-261.

Pope, R. (1991a). The Delusion of Certainty in Savage's Sure-Thing Principle. *Journal of Economic Psychology*, *12*, 209-241.

Pope, R. (1991b). Lowered welfare under the expected utility procedure. In A. Chikán (Ed.), *Progress in decision, utility and risk* (p. 125-133). Kluwer, Dordrecht.

Pope, R. (1995). Towards a more precise decision framework, a separation of the negative utility of chance from diminishing marginal utility and the preference for safety. *Theory and Decision*, *39*, 241-265.

Pope, R. (1996/7). Debates on the utility of chance: A look back to move forward. *Journal for Science of Research (Zeitschrift für Wissenschaftsforschung)*, *11/12*, 43-92. (reprinted in J. Götschl (Ed.), On the Dynamics of Modern, Complex and Democratic Systems, Theory and Decision Library, Series A: Philosophy and Methodology of the Social Sciences, Kluwer, Dordrecht, reprint of Pope 1996/7)

Pope, R. (2000). Reconciliation with the utility of chance by elaborated outcomes destroys the axiomatic basis of expected utility theory. *Theory and Decision*, *49*, 223-234.

Pope, R. (2001). Evidence of deliberate violations of dominance due to secondary satisfactions - attractions to chance. *Homo Economicus*, *15*, 47-76.

Pope, R. (2004). Biasses from omitted risk effects in standard gamble utilities. *Journal of Health Economics*, *23*, 695-735.

Pope, R. (2005a). The illusion of risk effects in stochastic industrial modelling. *International Journal of Systems and Management*, *21*.

Pope, R. (2005b). The riskless utility mapping of expected utility and all theories imposing the dominance principle: its inability to include loans, commitments even with fully described decision trees. In S. Schmidt U. Traub (Ed.), *Advances in Public Economics: Utility, Choice & Welfare: A Festschrift in Honour of Christian Seidl* (p. 289-327). Springer.

Pope, R. (forthcoming). The incompatibility of an insurance industry with a competitive market and with eut and most non-eut theories under any market structure: Health insurance and genetic data instead with skat, the stages of knowledge ahead theory. *Journal of Health Economics*.

Pope, R., Leopold-Wildburger, U., & Leitner, J. (2003). Eine experimentelle Studie zum Verhalten unter Terrorismusgefahr (An Experiment on Behaviour under Terrorism Danger). In E. Reiter (Ed.), *Jahrbuch für Internationale Sicherheitspolitik (Yearbook on International Relations)* (p. 801-810). Mittler and Sohn, Berlin.

Pope, R., Selten, R., Hagen, J. von, & Kube, S. (forthcoming). Experimental evidence on the benefits of eliminating exchange rate uncertainties and why expected utility theory causes economists to miss them. *Economic Notes, Review of Banking, Finance and Monetary Economics*.

Putnam, H. (2002). *The collapse of the fact/value distinction and other essays including the rosenthal lectures.* Harvard University Press, Cambridge, Massachusetts.

Quiggin, J. (1982). A theory of anticipated utility. *Journal of Economic Behavior and Organisation, 3*, 323-343.

Quiggin, J. (1993). *Generalised expected utility theory: The rank dependent model.* Kluwer, Boston.

Rabin, M. (2000). Risk aversion and expected-utility theory: A calibration theorem. *Econometrica, 5*, 1281-1292.

Ramsey, F. (1926). Truth and probability. reproduced in 1950. In R. B. Braithwaite (Ed.), *The foundations of mathematics and other logical essays* (p. 156-184). The Humanities Press, New York.

Ramsey, F. (1950). Truth and probability. In R. Braithwaite (Ed.), *The foundations of mathematics and other logical essays* (p. 156-184). The Humanities Press, New York.

Roberts, N. (2004). *The basis and ethics of terrorism.* (open letter to Dell Higgie, New Zealand Ambassador for Counter Terrorism, Ministry of foreign Affairs, Wellington, New Zealand)

Rossman, M., & Selden, L. (1979). *Note on indirect cardinal utility and intertemporal choice under uncertainty.* (Discussion Paper Series No. 48, Department of Economics, Columbia University, New York)

Samuelson, P. (1952a). Probability, utility, and the independence axiom. *Econometrica, 20*, 670-678.

Samuelson, P. (1952b). Contents of letter to Herman Wold reported in Wold, H., 1952, rejoinder to Savage and Samuelson. *Econometrica, 20*, 664.

Samuelson, P. (1964). Principles of efficiency: Discussion. *American Economic Review (Papers and Proceedings), 54*, 93-96.

Sauermann, H., & Selten, R. (1962). Anspruchsanpassungstheorie der Unternehmung. *Zeitschrift für die gesamte Staatswissenschaft, 118*, 577-597.

Savage, L. (1952a). Claim that EU is not for multiple choices. *Econometrica, 20(4)*, 663-664.

Savage, L. (1952b). Reply to Wold's rejoinder on limiting EU to single choices. *Econometrica, 20(4)*, 664.

Savage, L. (1954). *The foundations of statistics.* John Wiley, New York.

Schade, C., Kunreuther, H., & Kaas, K. (2004). *Probability neglect and concern in low probabilities and high stakes.* (Working Paper, Wharton Risk Management and Decision Processes)

Schneeweiß, H. (1968a). Die Unverträglichkeit von (m, s)-Prinzip und Dominanzprinzip. *Unternehmensforschung, 12*, 180-184.

Schneeweiß, H. (1968b). Note on two dominance principles in decision theory. *Unternehmensforschung (Operations research), 12*, 213-216.

Schneeweiß, H. (1973a). The (m, s) decision criterion on the class of normal distributions. *Inference and Decision, 1*, 23-30.

Schneeweiß, H. (1973b). On the consistency of classical decision criteria. *Inference and Decision, 1*, 31-44.

Schoemaker, P. (1982). The expected utility model: Its variants, purposes, evidence and limitations. *Journal of Economic Literature, 20*, 529-563.

Scitovsky, T. (1976). *The joyless economy: An inquiry into human satisfaction and consumer dissatisfaction.* Oxford University Press, Oxford.

Selten, R. (1998). Aspiration adaptation theory. *Journal of Mathematical Psychology, 42*, 191-214.

Sen, A. (1993). Internal consistency of choice. *Econometrica, 61*, 495-521.

Sen, A. (2002). Rationality and freedom. In *Harvard university press, cambridge.*

Simon, H. (1955). A behavioral model of rational choice. *Quarterly Journal of Economics, 69*, 99-118.

Simon, H. (1989). Rationality in psychology and economics. In R. Hogarth & M. Reder (Eds.), *Rational choice: The contrast between economics and psychology* (p. 25-40). University of Chicago Press, Chicago.

Simon, H. (1991a). *Models of my life.* Basic Books, USA.

Simon, H. (1991b). Problem formulation and alternative generation in the decision making process. In A. Chikán (Ed.), *Progress in decision, utility and risk* (p. 77-84). Kluwer, Dordrecht.

Slovic, P., Fischhoff, B., & Lichtenstein, S. (1982). Response mode, framing, and information-processing effects in risk assessment. In R. Hogarth (Ed.), *New directions for methodology of social behavioral science: No. 11. question framing and response consistency* (p. 21-36). San Francisco: Jossey-Bass.

Somerfield, M., & McCrae, R. (2000). Stress and coping research: methodological challenges, theoretical advances, and clinical applications. *American Psychologist, 55*, 620-625.

Starmer, C., & Sugden, R. (1991). Does the random-lottery incentive system elicit true preferences? *American Economic Review, 81*, 971-979.

Stiglitz, J., & Weiss, A. (1981). Credit rationing in imperfect markets. *American Economic Review, 71*, 393-410.

Sunstein, C. (2003). Terrorism and probability neglect. *Journal of Risk and Uncertainty, 26*, 121-136.

Tversky, A. (1967). Utility theory and the additivity analysis of risky choices. *Journal of Experimental Psychology, 75*, 27-36.

Tversky, A., & Kahneman, D. (1992). Advances in prospect theory: Cumulative representation of uncertainty. *Journal of Risk and Uncertainty, 5*, 297-323.

Vickrey, W. (1945). Measuring marginal utility by reactions to risk. *Econometrica, 13*, 319-333.

Vickrey, W. (1964). Principles of efficiency: Discussion. *American Economic Review (Papers and Proceedings), 54*, 88-92.

Walsh, V. (1996). *Rationality, allocation and reproduction.* Clarendon Press, Oxford.

Webster, L. (2004). *Overcoming war and empire by incentivizing justice.* (Paper presented to the European association for Evolutionary Political Economy, Rethymmon, Crete, November 5-7 2004)

Wold, H. (1952). Ordinal preferences or cardinal utility? *Econometrica, 20*, 661-664.

Wu, G. (1991). *Anxiety and decision making under delayed resolution of uncertainty.* (PhD Thesis in Decision Sciences, supervisor Howard Raiffa, Harvard University.)

Author Index

Subject Index

Lecture Notes in Economics and Mathematical Systems

For information about Vols. 1–495
please contact your bookseller or Springer-Verlag

Vol. 496: P. Weiß, Unemployment in Open Economies. XII, 226 pages. 2001.

Vol. 497: J. Inkmann, Conditional Moment Estimation of Nonlinear Equation Systems. VIII, 214 pages. 2001.

Vol. 498: M. Reutter, A Macroeconomic Model of West German Unemployment. X, 125 pages. 2001.

Vol. 499: A. Casajus, Focal Points in Framed Games. XI, 131 pages. 2001.

Vol. 500: F. Nardini, Technical Progress and Economic Growth. XVII, 191 pages. 2001.

Vol. 501: M. Fleischmann, Quantitative Models for Reverse Logistics. XI, 181 pages. 2001.

Vol. 502: N. Hadjisavvas, J. E. Martínez-Legaz, J.-P. Penot (Eds.), Generalized Convexity and Generalized Monotonicity. IX, 410 pages. 2001.

Vol. 503: A. Kirman, J.-B. Zimmermann (Eds.), Economics with Heterogenous Interacting Agents. VII, 343 pages. 2001.

Vol. 504: P.-Y. Moix (Ed.), The Measurement of Market Risk. XI, 272 pages. 2001.

Vol. 505: S. Voß, J. R. Daduna (Eds.), Computer-Aided Scheduling of Public Transport. XI, 466 pages. 2001.

Vol. 506: B. P. Kellerhals, Financial Pricing Models in Continuous Time and Kalman Filtering. XIV, 247 pages. 2001.

Vol. 507: M. Koksalan, S. Zionts, Multiple Criteria Decision Making in the New Millenium. XII, 481 pages. 2001.

Vol. 508: K. Neumann, C. Schwindt, J. Zimmermann, Project Scheduling with Time Windows and Scarce Resources. XI, 335 pages. 2002.

Vol. 509: D. Hornung, Investment, R&D, and Long Run Growth. XVI, 194 pages. 2002.

Vol. 510: A. S. Tangian, Constructing and Applying Objective Functions. XII, 582 pages. 2002.

Vol. 511: M. Külpmann, Stock Market Overreaction and Fundamental Valuation. IX, 198 pages. 2002.

Vol. 512: W.-B. Zhang, An Economic Theory of Cities.XI, 220 pages. 2002.

Vol. 513: K. Marti, Stochastic Optimization Techniques. VIII, 364 pages. 2002.

Vol. 514: S. Wang, Y. Xia, Portfolio and Asset Pricing. XII, 200 pages. 2002.

Vol. 515: G. Heisig, Planning Stability in Material Requirements Planning System. XII, 264 pages. 2002.

Vol. 516: B. Schmid, Pricing Credit Linked Financial Instruments. X, 246 pages. 2002.

Vol. 517: H. I. Meinhardt, Cooperative Decision Making in Common Pool Situations. VIII, 205 pages. 2002.

Vol. 518: S. Napel, Bilateral Bargaining. VIII, 188 pages. 2002.

Vol. 519: A. Klose, G. Speranza, L. N. Van Wassenhove (Eds.), Quantitative Approaches to Distribution Logistics and Supply Chain Management. XIII, 421 pages. 2002.

Vol. 520: B. Glaser, Efficiency versus Sustainability in Dynamic Decision Making. IX, 252 pages. 2002.

Vol. 521: R. Cowan, N. Jonard (Eds.), Heterogenous Agents, Interactions and Economic Performance. XIV, 339 pages. 2003.

Vol. 522: C. Neff, Corporate Finance, Innovation, and Strategic Competition. IX, 218 pages. 2003.

Vol. 523: W.-B. Zhang, A Theory of Interregional Dynamics. XI, 231 pages. 2003.

Vol. 524: M. Frölich, Programme Evaluation and Treatment Choise. VIII, 191 pages. 2003.

Vol. 525: S. Spinler, Capacity Reservation for Capital- Intensive Technologies. XVI, 139 pages. 2003.

Vol. 526: C. F. Daganzo, A Theory of Supply Chains. VIII, 123 pages. 2003.

Vol. 527: C. E. Metz, Information Dissemination in Currency Crises. XI, 231 pages. 2003.

Vol. 528: R. Stolletz, Performance Analysis and Optimization of Inbound Call Centers. X, 219 pages. 2003.

Vol. 529: W. Krabs, S. W. Pickl, Analysis, Controllability and Optimization of Time-Discrete Systems and Dynamical Games. XII, 187 pages. 2003.

Vol. 530: R. Wapler, Unemployment, Market Structure and Growth. XXVII, 207 pages. 2003.

Vol. 531: M. Gallegati, A. Kirman, M. Marsili (Eds.), The Complex Dynamics of Economic Interaction. XV, 402 pages, 2004.

Vol. 532: K. Marti, Y. Ermoliev, G. Pflug (Eds.), Dynamic Stochastic Optimization. VIII, 336 pages. 2004.

Vol. 533: G. Dudek, Collaborative Planning in Supply Chains. X, 234 pages. 2004.

Vol. 534: M. Runkel, Environmental and Resource Policy for Consumer Durables. X, 197 pages. 2004.

Vol. 535: X. Gandibleux, M. Sevaux, K. Sörensen, V. Tkindt (Eds.), Metaheuristics for Multiobjective Optimisation. IX, 249 pages. 2004.

Vol. 536: R. Brüggemann, Model Reduction Methods for Vector Autoregressive Processes. X, 218 pages. 2004.

Vol. 537: A. Esser, Pricing in (In)Complete Markets. XI, 122 pages, 2004.

Vol. 538: S. Kokot, The Econometrics of Sequential Trade Models. XI, 193 pages. 2004.

Vol. 539: N. Hautsch, Modelling Irregularly Spaced Financial Data. XII, 291 pages. 2004.

Vol. 540: H. Kraft, Optimal Portfolios with Stochastic Interest Rates and Defaultable Assets. X, 173 pages. 2004.

Vol. 541: G.-y. Chen, X. Huang, X. Yang, Vector Optimization. X, 306 pages. 2005.

Vol. 542: J. Lingens, Union Wage Bargaining and Economic Growth. XIII, 199 pages. 2004.

Vol. 543: C. Benkert, Default Risk in Bond and Credit Derivatives Markets. IX, 135 pages. 2004.

Vol. 544: B. Fleischmann, A. Klose, Distribution Logistics. X, 284 pages. 2004.

Vol. 545: R. Hafner, Stochastic Implied Volatility. XI, 229 pages. 2004.

Vol. 546: D. Quadt, Lot-Sizing and Scheduling for Flexible Flow Lines. XVIII, 227 pages. 2004.

Vol. 547: M. Wildi, Signal Extraction. XI, 279 pages. 2005.

Vol. 548: D. Kuhn, Generalized Bounds for Convex Multistage Stochastic Programs. XI, 190 pages. 2005.

Vol. 549: G. N. Krieg, Kanban-Controlled Manufacturing Systems. IX, 236 pages. 2005.

Vol. 550: T. Lux, S. Reitz, E. Samanidou, Nonlinear Dynamics and Heterogeneous Interacting Agents. XIII, 327 pages. 2005.

Vol. 551: J. Leskow, M. Puchet Anyul, L. F. Punzo, New Tools of Economic Dynamics. XIX, 392 pages. 2005.

Vol. 552: C. Suerie, Time Continuity in Discrete Time Models. XVIII, 229 pages. 2005.

Vol. 553: B. Mönch, Strategic Trading in Illiquid Markets. XIII, 116 pages. 2005.

Vol. 554: R. Foellmi, Consumption Structure and Macroeconomics. IX, 152 pages. 2005.

Vol. 555: J. Wenzelburger, Learning in Economic Systems with Expectations Feedback (planned) 2005.

Vol. 556: R. Branzei, D. Dimitrov, S. Tijs, Models in Cooperative Game Theory. VIII, 135 pages. 2005.

Vol. 557: S. Barbaro, Equity and Efficiency Considerations of Public Higer Education. XII, 128 pages. 2005.

Vol. 558: M. Faliva, M. G. Zoia, Topics in Dynamic Model Analysis. X, 144 pages. 2005.

Vol. 559: M. Schulmerich, Real Options Valuation. XVI, 357 pages. 2005.

Vol. 560: A. von Schemde, Index and Stability in Bimatrix Games. X, 151 pages. 2005.

Vol. 561: H. Bobzin, Principles of Network Economics. XX, 390 pages. 2006.

Vol. 562: T. Langenberg, Standardization and Expectations. IX, 132 pages. 2006.

Vol. 563: A. Seeger (Ed.), Recent Advances in Optimization. XI, 455 pages. 2006.

Vol. 564: P. Mathieu, B. Beaufils, O. Brandouy (Eds.), Artificial Economics. XIII, 237 pages. 2005.

Vol. 565: W. Lemke, Term Structure Modeling and Estimation in a State Space Framework. IX, 224 pages. 2006.

Vol. 566: M. Genser, A Structural Framework for the Pricing of Corporate Securities. XIX, 176 pages. 2006.

Vol. 567: A. Namatame, T. Kaizouji, Y. Aruga (Eds.), The Complex Networks of Economic Interactions. XI, 343 pages. 2006.

Vol. 568: M. Caliendo, Microeconometric Evaluation of Labour Market Policies. XVII, 258 pages. 2006.

Vol. 569: L. Neubecker, Strategic Competition in Oligopolies with Fluctuating Demand. IX, 233 pages. 2006.

Vol. 570: J. Woo, The Political Economy of Fiscal Policy. X, 169 pages. 2006.

Vol. 571: T. Herwig, Market-Conform Valuation of Options. VIII, 104 pages. 2006.

Vol. 572: M. F. Jäkel, Pensionomics. XII, 316 pages 2006.

Vol. 573: J. Emami Namini, International Trade and Multinational Activity, X, 159 pages, 2006.

Vol. 574: R. Kleber, Dynamic Inventory Management in Reverse Logisnes, XII, 181 pages, 2006.

Vol. 575: R. Hellermann, Capacity Options for Revenue Management, XV, 199 pages, 2006.

Vol. 576: J. Zajac, Economics Dynamics, Information and Equilibnum, X, 284 pages, 2006.

Vol. 577: K. Rudolph, Bargaining Power Effects in Financial Contracting, XVIII, 330 pages, 2006.

Vol. 578: J. Kühn, Optimal Risk-Return Trade-Offs of Commercial Banks, IX, 149 pages, 2006.

Vol. 579: D. Sondermann, Introduction to Stochastic Calculus for Finance, X, 136 pages, 2006.

Vol. 580: S. Seifert, Posted Price Offers in Internet Auction Markets, IX, 186 pages, 2006.

Vol. 581: K. Marti; Y. Ermoliev; M. Makowsk; G. Pflug (Eds.), Coping with Uncertainty, XIII, 330 pages, 2006.

Vol. 582: J. Andritzky, Sovereign Default Risks Valuation: Implications of Debt Crises and Bond Restructurings. VIII, 251 pages, 2006.

Vol. 583: I.V. Konnov, D.T. Luc, A.M. Rubinov (Eds.), Generalized Convexity and Related Topics, IX, 469 pages, 2006.

Vol. 584: C. Bruun, Adances in Artificial Economics: The Economy as a Complex Dynamic System. XVI, 296 pages, 2006.

Vol. 585: R. Pope, J. Leitner, U. Leopold-Wildburger, The Knowledge Ahead Approach to Risk, XVI, 218 pages, 2007.

Vol. 586: B. Lebreton, Strategic Closed-Loop Supply Chain Management. X, 150 pages, 2007.

Vol. 587: P.N. Baecker, Real Options and Intellectual Property: Capital Budgeting Under Imperfect Patent Protection. X, 276 pages, 2007 (planned).

Vol. 588: D. Grundel, M. Robert, P. Panos, O. Prokopyev (Eds.), Cooperative Systems: Control and Optimization. IX, 401 pages, 2007 (planned).

MIX
Papier aus verantwortungsvollen Quellen
Paper from responsible sources
FSC® C105338

If you have any concerns about our products,
you can contact us on
ProductSafety@springernature.com

In case Publisher is established outside the EU,
the EU authorized representative is:
Springer Nature Customer Service Center GmbH
Europaplatz 3, 69115 Heidelberg, Germany

Printed by Libri Plureos GmbH
in Hamburg, Germany